Letts study aids

Revise Chemistry

A complete revision course for O level and CSE

Bob McDuell BSc (Hons)

Deputy Headmaster, Berry Hill High School, Stoke-on-Trent

Charles Letts & Co Ltd
London, Edinburgh & New York

First published 1979
by Charles Letts & Co Ltd
Diary House, Borough Road, London SE1 1DW
Reprinted 1980
Revised 1981
Reprinted 1982

Design: Ben Sands
Illustrations: Rod Fraser

ISBN 0 85097 387 2

Printed in Great Britain by
Charles Letts (Scotland) Ltd

Section I

INTRODUCTION AND GUIDE TO USING THIS BOOK

Knowing how to prepare for an examination can be difficult. *Revise Chemistry* will help to make it easier by providing the GCE 'O' level, SCE and CSE Chemistry student with a complete revision course. This book contains not only all the necessary topics with test-yourself sections, but an analysis of actual past examination questions, skeleton answers, and some examination advice. To get the best from the study scheme in this book you are advised to follow the procedure outlined below.

(a) Using the table of analysis of examination syllabuses. Find the column containing the information about the syllabus you are studying. (Remember, many Examination Boards have alternative Chemistry syllabuses, so if you are in any doubt about the requirements of your particular Board please check with your teacher—the table is intended only as a useful guide.) In the column you will find the following information:

(i) the number of examination papers.
(ii) whether there is a practical examination. (The figure in brackets is the mark allocated to the practical examination.)
(iii) whether there is any teacher assessment. (This is a mark awarded by your teacher for the work you have done at school.)
(iv) whether fixed response and free response questions (see Section IV) are set.
(v) the Chemistry content of the GCE, SCE and CSE courses—the most important information! This has been divided into 42 units, but not all of these units need to be studied for any one syllabus. The key to the symbols in the columns is:

● unit required for a syllabus (blanks indicate unit not required)
○ unit optional for a syllabus
●/○ unit required as part of a syllabus but also required in greater depth as an option
* units are subdivided in the text—only part of a unit (A, B or C) may be required

You are advised to work through as many of the required units as possible as well as some of the optional units. Your teacher will advise you which options you should study. You may still find it useful to consult the Examination Board's syllabus as it may give valuable information in the form of additional notes. You can write to the Boards to buy syllabuses and past papers. A list of their addresses is given on page 4.

(b) Revising and testing yourself. The best advice that can be given to you, as an intending candidate, is to prepare thoroughly and to start the preparation in plenty of time. Studying must be more than simply reading through a book from start to finish. If a book is read in this way a number of times, it is likely that the first chapters may be well understood, but later chapters will be less familiar. Learning is best done in small regular portions. After you have worked through each unit, turn to the appropriate self-test (Section III). By using questions which highlight the difficulties and common errors, the tests are designed to see if you have mastered the unit. If you fail to get half marks in the test, go back and prepare more thoroughly.

(c) Examination technique. Section IV gives you hints for the examination and analyses the different types of question you are likely to meet.

(d) Practice in answering questions. In Section V there is a selection of examination questions for you to work through. Answers are given at the end of each group of questions.

Table of analysis of examination syllabuses

		SEB	AEB		Cambridge	JMB	London		Oxford		O and C	SUJB	WJEC	ALSEB	EAEB			Unit
	Level	O	O	O	O	O	O	O	O	O	O	O	O	CSE	CSE	CSE	CSE	
	Syllabus		1	2	5070			Nuffield	055	056					Joint	North	South	
	Number of papers	2	2	3	2	1	2	2	2	2	2	1	1	2	2	2	2	
	Practical examination				25%				$37\frac{1}{2}$%					15%		20%		
	Teacher assessment			○				○						25%	20%			
	Fixed response	45%	40%	25%	30%	40%	40%	●			●				40%	40%	45%	
	Free response	55%	60%	75%	45%	60%	60%	●	$62\frac{1}{2}$%	100%	●	100%	100%	60%	40%	40%	55%	
1	Elements, mixtures and compounds	●	●	●	●	●	●	●	●	●	●	●	●	●	●	●	●	1
2	Separation techniques	●	●	●	●	●	●	●	●	●	●	●	●	●	●	●	●	2
3	Symbols, formulae and equations	●	●	●	●	●	●	●	●	●	●	●	●	●	●	●	●	3
4	Structure of the atom	●	●	●	●	●	●	●	●	●	●	●	●	●	●	●	●	4
5	Bonding and structure	●	●	●	●	●	●	●	●	●	●	●	●	●	●	●	●	5
6	States of matter	●	●	●	●	●	●	●	●	●	●	●	●	●	●	●	●	6
7	Some laws relating to gases						●		●	●	●		●					7
8	Air and combustion	●	●	●	●	●	●	●	●	●	●		●	●	●	●	●	8
9	Oxygen and oxides	●	●	●	●	●	●	●	●	●	●	●	●	●	●	●	●	9
10	Hydrogen	●	●	●	●	●	●	●	●	●	●	●		●		●	●	1
11	Water	●	●	●	●	●	●	●/○	●	●	●	●	●	●/○	●	●	●	1
12	The mole and calculations	●	●	●	●	●	●	●	●	●	●	●	●	●	●	●	●	1
13	Laws of chemical combination*		B	●	B	●	B	B	●	B	●	●	●		●			1
14	Metals (I)	●	●	●	●	●	●	●/○	●	●	●	●	●	●/○	●	●	●	1
15	Metals (II)	●	●	●	●	●	●	●	●	●	●	●	●	●	●	●	●	1
16	Oxidation and reduction	●	●	●	●	●	●	●/○	●	●	●	●	●	●	●	●		1
17	Effect of electricity*	A	●	A	●	A	●	●	●	A	●	●	A	A	A	A	A	1
18	Chemical families*		●		●	●	●	●/○	●	●	●			●	●	●	●	1
19	Acids, bases and salts	●	●	●	●	●	●	●/○	●	●	●	●	●	●	●	●	●	1
20	Rates of reaction		●	●	●	●	●	●	●	●	●		●	●	●	●	●	2
21	Reversible reactions		●	●			●	●	●	●			●		●			2
22	Sulphur	●	●	●	●	●	●	●	●	●	●	●	●	●/○		●		2
23	Sulphides and hydrogen sulphide			●					●		●	●		●/○				2
24	Oxides of sulphur	●	●	●	●	●	●		●	●	●	●	●	●/○				2
25	Sulphuric acid and sulphates	●	●	●	●	●	●		●	●	●	●	●	●		●	●	2
26	Nitrogen	●		●	●	●			●	●	●	●		●		●		2
27	Ammonia and ammonium compounds	●	●	●	●	●	●	●	●	●	●	●	●	●	●	●	●	2
28	Nitric acid and nitrates	●	●	●	●	●	●	●	●	●	●	●	●	●				2
29	Oxides of nitrogen		●									●						2
30	Agricultural chemistry	●		●				●							●	●		3
31	Phosphorus						●		●									3
32	Hydrogen chloride and chlorine*		●	●	●	●	●	A	●	●	●	●	●	●			B	3
33	Carbon and organic chemistry*	●	●	●	●	●	●	A, C	●	●	●	●	●	A, C	A, C	A, C	A, C	3
34	Carbon dioxide and carbonates	●	●	●	●	●	●	●	●	●	●	●	●	●	●	●	●	3
35	Polymers	●	●	●	●	●	●	●	●	●	●	●	●	○	●	●	●	3
36	Fuels	●		●		●	●	●						○	●	●		3
37	Energy in Chemistry	●	●	●		●	●	●/○	●	●	●	●	●	●	●	●	●	3
38	Social and economic considerations	●		●		●				●				○		●		3
39	Radioactivity			●				●		●			●	○		●		3
40	Quantitative volumetric chemistry	●			●	●	●	○	●		●		●			●		4
41	Qualitative chemistry	●			●	●	●		●		●	●				●	●	4
42	The diaphragm cell					●											●	4

EMREB		LREB				NREB	NWREB	SREB		SEREB		SWEB	WJEC	WMEB		WY & LREB	YREB	JMB/WMEB	Unit
CSE	CSE	CSE	CSE	CSE	CSE	CSE	CSE	CSE	CSE	CSE	CSE	CSE	CSE	CSE	CSE	CSE	CSE	16+ (CSE/O)	
1	2	A	B	C	D			R(A)	R(B)		Nuffield			A	B				
2	2	2	2	2	2	1	1	1	1	2	2	2	1	2	2	1	2	2	
●	●	20%	20%				30%					20%							
				20%	20%	20%		10%	10%	●	●			20%	20%	15%	●	25%	
	●	30%	30%	30%	30%		30%	40%	40%	●	●			●	●		●	16%	
70%	●	50%	50%	50%	50%	80%	40%	50%	50%	●	●	80%	100%	●	●	85%	●	59%	
●	●	●	●	●	●	●	●	●	●	●	●	●	●	●	●	●	●	●	1
●	●	●	●	●	●	●	●	●	●	●	●	●	●	●	●	●	●	●	2
●	●	●	●	●	●	●	●	●	●	●	●	●	●	●	●	●	●	●	3
●	●	●	●	●	●	●	●	●	●	●	●	●	●	●	●	●	●	●	4
●	●/○	●	●	●	●	●	●	●	●	●	●	●	●	●	●		●	●	5
●/○	●	●	●	●	●	●	●	●	●	●	●	●	●	●	●	●	●	●	6
																			7
●	●	●	●	●	●	●	●	●	●	●	●	●	●	●	●	●	●	●	8
●	●	●	●	●	●	●	●	●	●	●	●	●	●	●	●	●	●	●	9
●	●	●	●	●	●	●		●	●	●		●	●	●	●	●	●	●	10
●	●	●	●	●	●	●	●	●		●	●	●	●	●	●	●	●	●	11
●/○	●	●	●	●	●	●	●	●	●	●	●	●	●	●	●	●	●	●	12
A								A	B					A	A				13
●/○	●	●	●	●	●	●	●	●	●	●	●	●	●	○	●	●	●	○	14
●	●	●	●	●	●	●	●	●	●	●	●	●	●	●	●	●	●	○	15
●	●	●	●	●	●	●	●	●	●	●	●	●	●	●	●	●	●	●	16
A	●/○	A	A	A	A	A	A	A	●	A	A	A	A	A	A	A	A	A	17
●/○	A	●	●	●	●	●	●	A	●		●	●	●			●	●	●	18
●	●	●	●	●	●	●	●	●	●	●	●	●	●	●	●	●	●	●	19
●	●	●	●	●	●	●	●		●		●	●		●	●	●	●	○	20
	●	●	●	●	●				●		●	●					●	○	21
●/○		●		●		●	●	●		●		●		○			●	○	22
								●		●				○				○	23
●		●		●				●		●		●	●	○			●	○	24
●/○		●		●		●	●	●		●		●	●	○		●	●	○	25
●	●					●		●		●	●		●	○	●	●	●	○	26
●/○	●	●	●	●	●	●	●	●	●	●	●	●	●	○	●	●	●	○	27
●		●	●	●	●		●	●		●		●	●	○	●	●	●	○	28
														○				○	29
			●		●			●			●			●	●				30
														●					31
●/○	A	●	A	●	A	B	●	●	A	●	A	●	B	○			●	○	32
A, C	C	●	A, C	●	A, C	A, C	A, C	●	A, C	●	A, C	A, C	●	○	A, C	●	●	○	33
●	●	●	●	●	●	●	●	●	●	●	●	●	●	●	●	●	●	●	34
●	○	●	●	●	●	●	●	●	●	●	●	●	●	○	●	●	●	○	35
●	○	●	●	●	●		●	●	●			●	●		●	●			36
●	●/○		●		●	●			●	●	●	●		●	●	●	●	●	37
		●		●		●		●		●					●	●			38
		●		●				●	●		●		●						39
		●	●	●	●		●				●								40
●	●	●	●	●	●	●	●	●		●		●		●	●		●	●	41
		●		●															42

Units are subdivided in the text – only part of a unit (A, B or C) is required.

EXAMINING BOARDS

GCE Boards

AEB	Associated Examining Board Wellington House, Aldershot, Hampshire GU11 1BQ
Cambridge	University of Cambridge Local Examinations Syndicate Syndicate Buildings, 17 Harvey Road, Cambridge CB1 2EU
JMB	Joint Matriculation Board Manchester M15 6EU
London	University Entrance and School Examinations Council University of London, 66–72 Gower Street, London WC1E 6E
Oxford	Oxford Local Examinations Delegacy of Local Examinations, Ewert Place, Summertown, Oxford OX2 7BX
O and C	Oxford and Cambridge Schools Examination Board 10 Trumpington Street, Cambridge; and Elsfield Way, Oxford
SUJB	Southern Universities' Joint Board for School Examinations Cotham Road, Bristol BS6 6DD
WJEC	Welsh Joint Education Committee 245 Western Avenue, Cardiff CF5 2YX

SCE Board

SEB	Scottish Examination Board Ironmills Road, Dalkeith, Midlothian EH22 1BR

CSE Boards

ALSEB	Associated Lancashire Schools Examining Board 17 Harter Street, Manchester M1 6HL
EAEB	East Anglian Examinations Board The Lindens Lexden Road, Colchester, Essex CO3 3RL
EMREB	East Midland Regional Examinations Board Robins Wood House, Robins Wood Road, Apsley, Nottingham NG8 3NH
LREB	London Regional Examinations Board Lyon House, 104 Wandsworth High Street, London SW18 4LF
NREB	North Regional Examinations Board Wheatfield Road, Westerhope, Newcastle upon Tyne NE5 5JZ
NWREB	North West Regional Examinations Board Orbit House, Albert Street, Eccles, Manchester M30 0WL
SREB	Southern Regional Examinations Board 53 London Road, Southampton SO9 4YL
SEREB	South East Regional Examinations Board Beloe House, 2/4 Mount Ephraim Road, Royal Tunbridge Wells, Kent TN1 1EU
SWEB	South Western Examinations Board 23–29 Marsh Street, Bristol BS1 4BP
WJEC	Welsh Joint Education Committee 245 Western Avenue, Cardiff CF5 2YX
WMEB	West Midlands Examination Board Norfolk House, Smallbrook Queensway, Birmingham B5 4NJ
WY & LREB	West Yorkshire and Lindsey Regional Examining Board Scarsdale House, 136 Derbyshire Lane, Sheffield S8 8SE
YREB	Yorkshire Regional Examinations Board 31–33 Springfield Avenue, Harrogate, North Yorkshire HG1 2HW

Section II Core units 1–42

1 Elements, mixtures and compounds

1.1 ELEMENTS

An **element** is a pure substance which cannot be split up by chemical reaction.

There are 105 chemical elements; most of these occur naturally but a few are man-made. Most of the elements are *solid* and *metallic*. However, bromine and mercury are *liquid* at room temperature and pressure, while hydrogen, helium, nitrogen, oxygen, fluorine, neon, chlorine, argon, krypton, xenon and radon are *gases* at room temperature and pressure.

Elements that have been mixed together can still be separated. For example, a mixture of iron and sulphur can be separated with a magnet.

1.2 COMPOUNDS

Certain mixtures of elements react together or combine (usually when heated) to form compounds. The formation of a compound from its constituent elements is called **synthesis**, and energy is usually liberated or released during this process (see 23.1). The compounds thus formed have very different properties from the elements combining together. Splitting up a compound into its constituent elements is not an easy process. The proportions of the different elements in a compound are fixed.

1.3 EXAMPLES OF COMPOUND FORMATION

(*i*) When a mixture of powdered iron and sulphur is heated, a reaction takes place with the evolution of energy forming iron(II) sulphide.

$$\text{Fe(s)} + \text{S(s)} \rightarrow \text{FeS(s)}$$
$$\text{iron} + \text{sulphur} \rightarrow \text{iron(II) sulphide}$$

(*ii*) When a mixture of the gaseous elements hydrogen and oxygen is exploded, a reaction takes place forming water (a liquid). The properties of water are different from the properties of hydrogen and oxygen.

$$2H_2(g) \times O_2(g) \rightarrow 2H_2O(1)$$
$$\text{hydrogen} + \text{oxygen} \rightarrow \text{water}$$

1.4 NAMING COMPOUNDS

(*i*) Compounds ending in **-ide** contain two elements.
E.g. copper(II) oxide is a compound of copper and oxygen;
calcium chloride is a compound of calcium and chlorine.
(Exceptions are compounds such as sodium hydroxide which contains the elements sodium, hydrogen and oxygen.)

(*ii*) Compounds ending in **-ate** or **-ite** contain oxygen. There is a greater proportion of oxygen in the compound ending in -ate.
E.g. sodium sulphate Na_2SO_4
sodium sulphite Na_2SO_3

(*iii*) Compounds with a prefix **per-** contain extra oxygen.
E.g. sodium oxide Na_2O
sodium peroxide Na_2O_2

(*iv*) Compounds with a prefix **thio-** contain a sulphur atom in place of an oxygen atom.
E.g. sodium sulphate Na_2SO_4
sodium thiosulphate $Na_2S_2O_3$

2 Separation techniques in chemistry

A chemist is frequently involved in separating mixtures or purifying substances. There are a number of methods available as outlined below.

2.1 Separating insoluble impurities from a soluble substance

E.g. removing sand and other impurities from salt solution

When crushed rock salt is added to water, the salt **dissolves** and forms a salt **solution**. The sand and other impurities remain undissolved, and this material can be removed by **filtration** (Fig. 2.1).

The salt solution passing through the filter paper contains no solid impurities. In order to obtain a sample of pure salt, the salt solution is **evaporated** (Fig. 2.2).

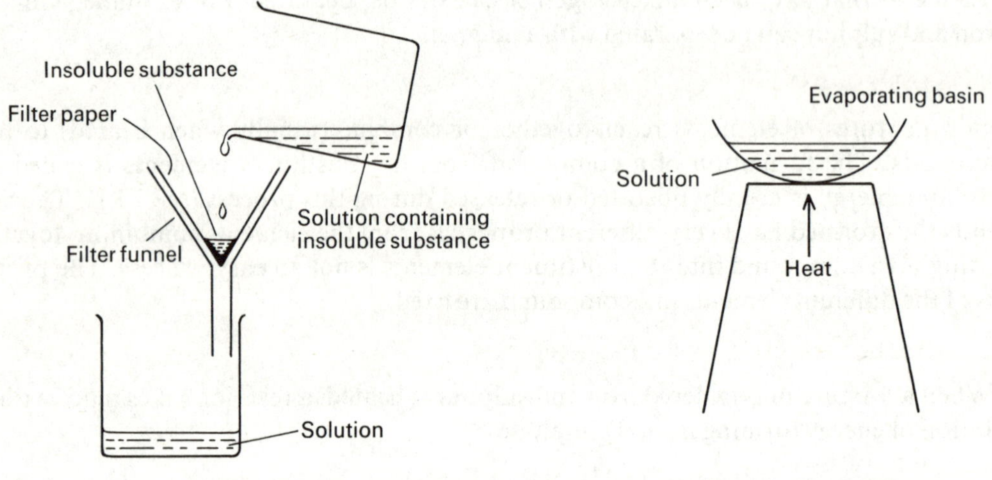

Fig. 2.1 Filtration

Fig. 2.2 Evaporation

Undissolved material can also be removed by **centrifuging**. This is used when small quantities of material are used, *e.g.* blood samples.

2.2 Separating a liquid from a solution of a solid in a liquid

E.g. producing pure water (distilled water) from sea water

This process is called **distillation**. When the flask is heated the solution boils and steam passes into the condenser. In the condenser, the steam is cooled by cold water passing through the outer condenser tube. The steam **condenses** and the **distillate** (distilled water) collects in the receiver. The impurities are concentrated in the flask (Fig. 2.3).

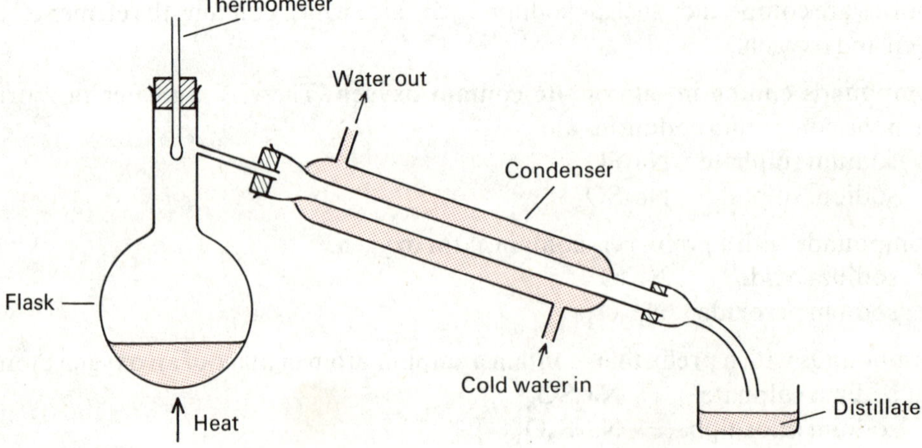

Fig. 2.3 Distillation

2.3 SEPARATING A LIQUID FROM A MIXTURE OF MISCIBLE LIQUIDS

E.g. ethanol from a mixture of ethanol and water

(Miscible liquids are liquids that mix together completely to form a single layer.)

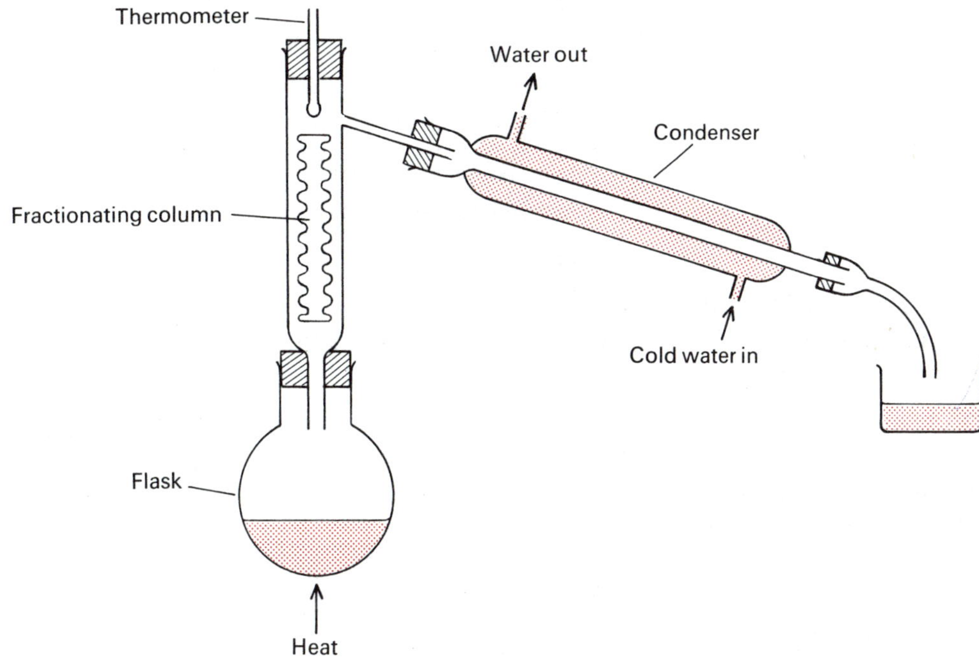

Fig. 2.4 Fractional distillation

This process is called **fractional distillation** (Fig. 2.4). It relies on the difference in boiling point of the two liquids (*e.g.* water 100°C, ethanol 78°C).

When the flask is heated the ethanol boils more readily than the water. The water (with the higher boiling point) condenses in the fractionating column and drips back into the flask. The ethanol distils over first. When all the ethanol has distilled over, the temperature (recorded on the thermometer) rises, and water distils over and is collected in a different receiver.

2.4 SEPARATING A MIXTURE OF IMMISCIBLE LIQUIDS

*E.g. a mixture of water and hexane**

These two liquids are **immiscible** and form two separate layers. The hexane layer forms above the water (or aqueous layer), because water is denser than hexane. These two liquids could be separated using a tap funnel.

2.5 SEPARATING A MIXTURE OF SIMILAR COMPOUNDS IN SOLUTION

E.g. separating the dyes present in a sample of ink

This process is called **chromatography**. It is a very sensitive method that can be used to separate similar compounds in solution.

If a spot of dye solution is put onto a filter paper and the spot enlarged by slowly dropping solvent onto the centre of the spot, the different components of the dye spread out at different rates. Each component forms a definite ring on the filter paper.

Chromatography experiments are often carried out using square sheets of filter paper. Spots of dye solutions are put along the base line of a sheet of filter paper. The filter paper is coiled into a cylinder and the cylinder is put into a tank containing a small volume of solvent. The lid is put on the tank and the solvent slowly rises up the filter paper. When the solvent has nearly reached the top of the filter paper, the cylinder is removed and the position of the solvent is marked.

*Carbon tetrachloride is not now recommended for use in schools.

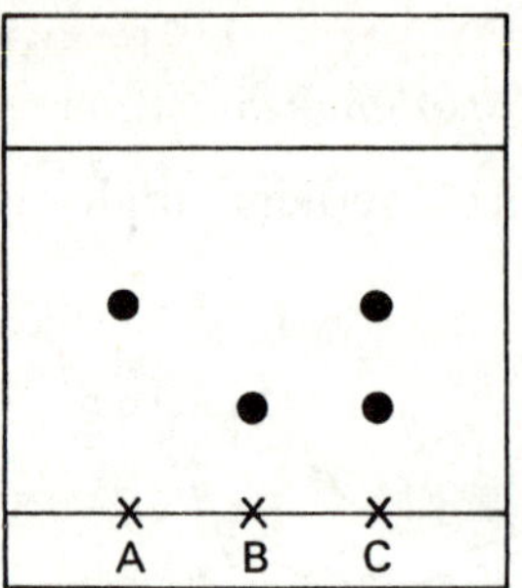

Fig. 2.5 A chromatogram

In Fig. 2.5, dyes A and B are either pure substances or a mixture of dyes not separated with the solvent used. Dye C is composed of a mixture of A and B, because the original spot has separated into two spots corresponding to A and B.

Chromatography was originally devised to separate coloured substances in solution. It can, however, be used to separate colourless substances in solution, which can then be seen by spraying or dipping the filter paper into a suitable chemical (called a locating agent) which colours the spots produced.

2.6 SEPARATING A SOLID WHICH SUBLIMES FROM A SOLID WHICH DOES NOT SUBLIME

E.g. separating ammonium chloride (which sublimes) from sodium chloride (which does not sublime)

A substance is said to **sublime** if, on cooling, its vapour changes directly from vapour to solid without going through an intermediate liquid state. Usually a substance which sublimes also changes from solid to vapour without melting to a liquid.

If a mixture of ammonium chloride and sodium chloride is heated, the ammonium chloride turns directly to a vapour but the sodium chloride remains unchanged. When the vapour is cooled, solid ammonium chloride collects free from sodium chloride.

2.7 RECOGNISING A PURE SUBSTANCE

A **pure substance** has a **definite melting point**. The presence of an impurity **lowers the melting point** but also causes the substance to melt over a **range of temperature**. Calcium chloride is used to lower the melting point of sodium chloride in the extraction of sodium (see 15.5).

The boiling point of a substance depends on pressure. At atmospheric pressure, the boiling of a pure substance takes place at a particular temperature called the boiling point. The presence of dissolved impurities **increases the boiling point** slightly.

---

3 Symbols, formulae and equations

3.1 SYMBOLS

Each element is represented by a symbol. This is either one or two letters and is a shorthand that all chemists understand:

e.g.

Element	Symbol	
Hydrogen	H	The first letter of the name
Carbon	C	
Calcium	Ca	The first two letters of the name
Helium	He	
Chlorine	Cl	The first letter and one other letter in the name
Magnesium	Mg	
Iron	Fe	Two letters not coming from the name
Sodium	Na	

(N.B. The first letter is a capital letter and the second letter is a small letter.) A full list of the elements and their symbols will be found in Fig. 18.2 page 67.

3.2 FORMULAE

Each compound is represented by a formula which gives the proportions of the different elements in the compound by mass. The formula of any compound could be found by carrying out a suitable experiment. The formulae of many compounds can be found by use of the list of ions in Table 3.1.

Table 3.1 List of common ions

Positive ions		*Negative ions*	
Sodium	Na^+	Chloride	Cl^-
Potassium	K^+	Bromide	Br^-
Silver	Ag^+	Iodide	I^-
Copper(II)	Cu^{2+}	Hydroxide	OH^-
Lead	Pb^{2+}	Nitrate	NO_3^-
Magnesium	Mg^{2+}	Nitrite	NO_2^-
Calcium	Ca^{2+}	Hydrogencarbonate	HCO_3^-
Zinc	Zn^{2+}	Sulphate	SO_4^{2-}
Barium	Ba^{2+}	Sulphite	SO_3^{2-}
Iron(II)	Fe^{2+}	Carbonate	CO_3^{2-}
Iron(III)	Fe^{3+}	Oxide	O^{2-}
Aluminium	Al^{3+}	Sulphide	S^{2-}
Ammonium	NH_4^+	Phosphate	PO_4^{3-}
Hydrogen	H^+		

In forming the compound the number of ions used is such that the number of positive charges equals the number of negative charges:

e.g. sodium chloride is made up from Na^+ and Cl^- ions. Since a sodium ion has a single positive charge and a chloride ion has a single negative charge, the formula of sodium chloride is NaCl.

Sodium sulphate is made up from Na^+ and SO_4^2 iuns. Twicc as many sodium ions as sulphate ions are necessary in order to have equal numbers of positive and negative charges. The formula of sodium sulphate is Na_2SO_4. Table 3.2 contains further examples.

Table 3.2 Further examples of formulae

Compound	*Ions present*	*Formula*
Copper(II) oxide	$Cu^{2+}O^{2-}$	CuO
Ammonium chloride	$NH_4^+Cl^-$	NH_4Cl
Silver nitrate	$Ag^+NO_3^-$	$AgNO_3$
Magnesium chloride	$Mg^{2+}Cl^-$	$MgCl_2$
Magnesium hydroxide	$Mg^{2+}OH^-$	$Mg(OH)_2$
Aluminium nitrate	$Al^{3+}NO_3^-$	$Al(NO_3)_3$
Aluminium oxide	$Al^{3+}O^{2-}$	Al_2O_3
Hydrochloric acid	H^+Cl^-	HCl
Sulphuric acid	$H^+SO_4^2$	H_2SO_4
Nitric acid	$N^+NO_3^-$	HNO_3

N.B. (*i*) Acids contain H^+ ions.

(*ii*) A small number after a bracket multiplies everything inside the bracket. *E.g.* $Mg(OH)_2$ is composed of one magnesium, two oxygen and two hydrogen atoms. These are formed into three ions—one Mg^{2+} ion and two OH^- ions.

All of the compounds above are composed of ions. Many compounds are not ionised. The formulae of some of these compounds are shown in Table 3.3.

Table 3.3 Formulae of some common compounds

Compound	*Formula*	*Compound*	*Formula*
Water	H_2O	Sulphur dioxide	SO_2
Carbon dioxide	CO_2	Sulphur trioxide	SO_3
Carbon monoxide	CO	Ammonia	NH_3
Nitrogen monoxide	NO	Hydrogen chloride	HCl
Nitrogen dioxide	NO_2	Methane	CH_4

3.3 CHEMICAL EQUATIONS

Chemical equations are widely used in textbooks and examination papers. An equation is a useful summary of a chemical reaction, and it is always theoretically possible to obtain an equation from the results of an experiment. It is advisable, however, to be able to write important equations in an examination.

The steps in writing a chemical equation are as follows:

(*i*) Write down the equation as a word equation either using the information given or your memory. Include all the reacting substances and products.
E.g. calcium hydroxide + hydrochloric acid → calcium chloride + water

(*ii*) Fill in the correct formulae for all the reacting substances and products.

$$Ca(OH)_2 + HCl \rightarrow CaCl_2 + H_2O$$

(*iii*) The equation should then be balanced. During any chemical reaction, atoms cannot be created or destroyed (Law of conservation of mass, see 13.1). There must be the same number of atoms before and after the reaction. Only the proportions of the reacting substances and products can be altered—not the formulae.

$$Ca(OH)_2 + 2HCl \rightarrow CaCl_2 + 2H_2O$$

(*iv*) Finally, the states of reacting substances and products can be included in small brackets after the formulae. Thus:

- (s) for solid—sometimes (c) is seen for representing a crystalline solid;
- (l) for liquid;
- (g) for gas;
- (aq) for a solution with water as solvent.

These state symbols are not given or expected by all examination boards.

$$Ca(OH)_2(aq) + 2HCl(aq) \rightarrow CaCl(aq) + 2H_2O(l)$$

3.4 INFORMATION PROVIDED BY AN EQUATION

$$2Mg(s) + O_2(g) \rightarrow 2MgO(s)$$

magnesium + oxygen → magnesium oxide

This equation gives the following information:

2 moles of magnesium atoms (48 g) combine with 1 mole of oxygen molecules to produce 2 moles of magnesium oxide (80 g). It also tells us that magnesium and magnesium oxide are solids and oxygen is a gas.

A chemical equation gives:
(*i*) the chemicals reacting together (called reactants) and the chemicals produced (called products);
(*ii*) the physical states of reactants and products;
(*iii*) the quantities of chemicals reacting together and produced (see 12.10).

The equation does not, however, give information about energy changes in the reaction, the rate of the reaction and the conditions necessary for the reaction to take place.

3.5 IONIC EQUATIONS

Ionic equations are useful because they emphasise the important changes taking place in a chemical reaction. For example, the equation for the neutralisation reaction between sodium hydroxide and hydrochloric acid is:

$$NaOH(aq) + HCl(aq) \rightarrow NaCl(aq) + H_2O(l)$$
sodium hydroxide + hydrochloric acid → sodium chloride + water

Since all of the reactants and products (except water) are composed of ions, this equation could be written:

$$Na^+(aq)OH^-(aq) + H^+(aq)Cl^-(aq) \rightarrow Na^+(aq)Cl^-(aq) + H_2O\,(l)$$

An equation should show change and therefore anything present before and after the reaction can be deleted. The simplest ionic equation, deleting $Na^+(aq)$, and $Cl^-(aq)$, is therefore:

$$OH^-(aq) + H^+(aq) \rightarrow H_2O(l)$$

The same ionic equation also applies to similar reactions *e.g.* calcium hydroxide and nitric acid.

In addition to balancing in the usual way, the sum of the charges on the left hand side must equal the sum of the charges on the right hand side.

Other common ionic equations include:

$$2Fe^{2+}(aq) + Cl_2(g) \rightarrow 2Fe^{3+}(aq) + 2Cl^-(aq)$$ (see 14.8)

$$2Br^-(aq) + Cl_2(g) \rightarrow Br_2(aq) + 2Cl^-(aq)$$ (see 18.3)

$$Zn(s) + 2H^+(aq) \rightarrow Zn^{2+}(aq) + H_2(g)$$ (see 10.1)

4 Structure of the atom

4.1 PARTICLES IN ATOMS

All elements are made up from **atoms**. An atom is the smallest part of an element that can exist.

It has been found that the atoms of all elements are made up from three basic particles, and that the atoms of different elements contain different numbers of these three particles. These particles are:

proton	p	mass 1 a.m.u.	charge + 1	
electron	e	mass $\frac{1}{1800}$ a.m.u. (negligible)	charge − 1	(a.m.u. = atomic mass unit)
neutron	n	mass 1 a.m.u.	neutral	

Because an atom has no overall charge, the number of protons in any atom is equal to the number of electrons.

Atomic number. The atomic number is the number of protons in an atom.

Mass number. The mass number is the total number of protons and neutrons in an atom.

E.g. The mass number of carbon-12 is 12, and the atomic number is 6. Therefore a carbon-12 atom contains 6 protons (*i.e.* atomic number), 6 electrons, and 6 neutrons. This sometimes written as $^{12}_{6}C$ (the atomic number is written under the mass number).

Sodium-23 Mass number 23 Atomic number 11 (*i.e.* $^{23}_{11}Na$)
$p = 11, e = 11, n = 23 - 11 = 12.$

It is possible, with many elements, to get more than one type of atom. For example, there are three types of oxygen atom:

Oxygen-16	8p, 8e, 8n
Oxygen-17	8p, 8e, 9n
Oxygen-18	8p, 8e, 10n

These different types of atom of the same element are called **isotopes**. They are different because they contain different numbers of neutrons. (If they did not contain the same number of protons and the same number of electrons they would not be isotopes of oxygen.) Isotopes of the same element have similar chemical properties but slightly different physical properties.

There are two isotopes of chlorine—chlorine-35 and chlorine-37. An ordinary sample of chlorine contains approximately 75% chlorine-35 and 25% chlorine-37. This explains the fact that the relative atomic mass of chlorine is approximately 35.5. (The relative atomic mass of an element is the mass of an 'average atom' compared with the mass of a $^{12}_{6}C$ carbon atom.)

4.2 Arrangement of particles in an atom

The protons and neutrons are tightly packed together in the **nucleus** of an atom. The electrons move rapidly around the nucleus in distinct energy levels. Each energy level is capable of accommodating only a certain number of electrons. This is represented in a simplified form in Fig. 4.1.

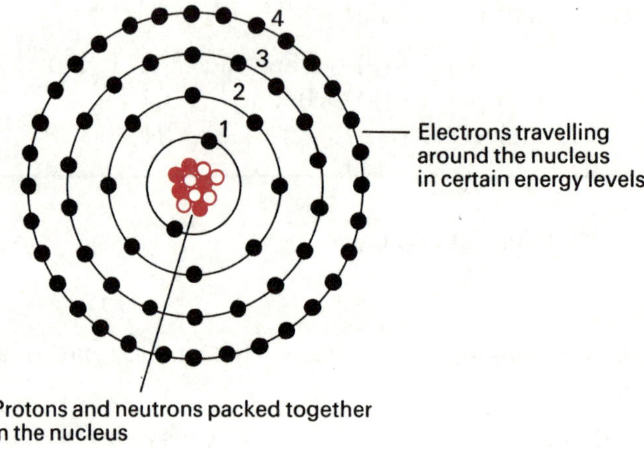

Fig. 4.1 Arrangement of particles in an atom

1. The first energy level (sometimes called the K shell and labelled 1 in Fig. 4.1) can only hold two electrons. This energy level is filled first.

2. The second energy level (sometimes called the L shell and labelled 2 in Fig. 4.1) can only hold eight electrons. This energy level is filled after the first energy level and before the third energy level.

3. The third energy level (sometimes called the M shell and labelled 3 in Fig. 4.1) can hold a maximum of 18 electrons. However, when eight electrons are in the third energy level there is a degree of stability and the next two electrons added go into the fourth energy level. Then extra electrons enter the third energy level until it contains a maximum of 18 electrons.

4. There are further energy levels, each containing a larger number of electrons than the preceeding energy level.

Table 4.1 gives the number of protons, neutrons and electrons in the principal isotopes of the first 20 elements. The electronic structure 2,8,1 denotes 2 electrons in the 1st energy level, 8 in the second, and 1 in the third. This is sometimes called the electron configuration of an atom.

Table 4.1 Numbers of protons, neutrons and electrons in the principal isotopes of the first 20 elements

Element	*Atomic number*	*Mass number*	*Number of* p	n	e	*Arrangement of electrons*
Hydrogen	1	1	1	0	1	1
Helium	2	4	2	2	2	2
Lithium	3	7	3	4	3	2,1
Beryllium	4	9	4	5	4	2,2
Boron	5	11	5	6	5	2,3
Carbon	6	12	6	6	6	2,4
Nitrogen	7	14	7	7	7	2,5
Oxygen	8	16	8	8	8	2,6
Fluorine	9	19	9	10	9	2,7
Neon	10	20	10	10	10	2,8
Sodium	11	23	11	12	11	2,8,1
Magnesium	12	24	12	12	12	2,8,2
Aluminium	13	27	13	14	13	2,8,3
Silicon	14	28	14	14	14	2,8,4
Phosphorus	15	31	15	16	15	2,8,5
Sulphur	16	32	16	16	16	2,8,6
Chlorine	17	35	17	18	17	2,8,7
Argon	18	40	18	22	18	2,8,8
Potassium	19	39	19	20	19	2,8,8,1
Calcium	20	40	20	20	20	2,8,8,2

5 Bonding and structure

In Unit 4 the arrangement of protons, neutrons and electrons within an atom was explained. In this unit we are concerned with the ways in which atoms join together (called **bonding**), and the arrangements of particles produced (called **structure**). There are five methods of bonding which will be discussed below.

5.1 Ionic (or electrovalent) bonding

This involves a **complete transfer of electrons** from one atom to another. Two examples are given below:

(*a*) *sodium chloride*
A sodium atom has an electronic structure of 2,8,1 (*i.e.* one more electron than the stable inert gas electronic arrangement of 2,8). A chlorine atom has an electronic arrangement of 2,8,7 (*i.e.* one electron less than the stable electronic arrangement 2,8,8).

If each sodium atom loses one electron (and forms a sodium ion Na^+) and each chlorine atom gains one electron (and forms a chloride ion Cl^-), both the ions formed have **stable electronic arrangements.** The ions are held together by strong electrostatic forces.

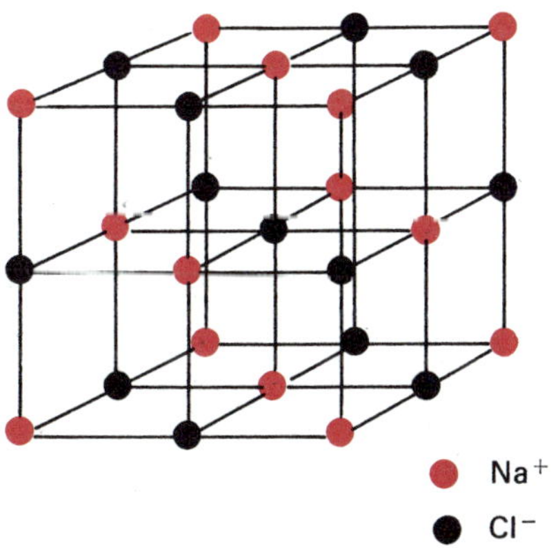

Fig. 5.1 Structure of sodium chloride

It is incorrect to speak of a 'sodium chloride *molecule*'. A sodium chloride *crystal* consists of a regular arrangement of equal numbers of sodium and chloride ions. This is called a **lattice** (Fig. 5.1).

(*b*) *magnesium oxide*
Electronic arrangement in magnesium atom 2,8,2
Electronic arrangement in oxygen atom 2,6
Two electrons are lost by each magnesium atom to form Mg^{2+} ions. Two electrons are gained by each oxygen atom to form O^{2-} ions.

Loss of one or two electrons by a metal during ionic bonding is common, *e.g.* NaCl or MgO. If three electrons are lost by a metal the resulting compound shows some covalent character, *e.g.* $AlCl_3$.

5.2 COVALENT BONDING

Covalent bonding involves the **sharing of electrons** rather than complete transfer. Two examples are given below:

(*a*) *chlorine molecule* (Cl_2)
A chlorine atom has an electronic arrangement of 2,8,7. When two chlorine atoms bond together they form a chlorine molecule. If one electron was transferred from one chlorine atom to the other, only one atom could achieve a stable electronic arrangement.

One electron from each atom is donated to form a pair of electrons which is shared between both atoms holding them together. This is called a single covalent bond. Fig. 5.2 shows a simple representation of a chlorine molecule. This is often shown as Cl—Cl.

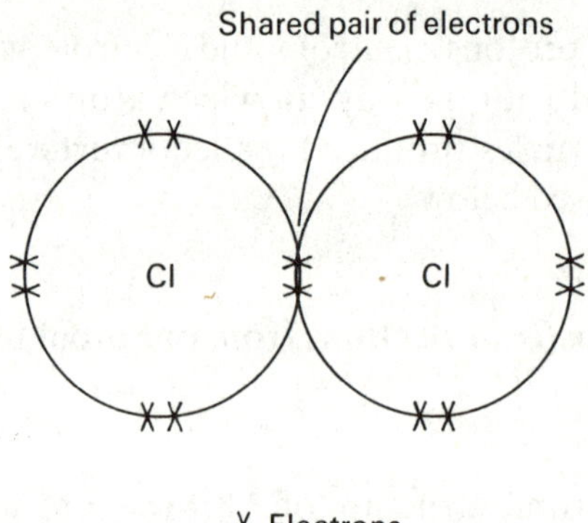

Fig. 5.2 Chlorine molecule

(*b*) *oxygen molecule* (O_2)
An oxygen atom has an electronic arrangement of 2,6. In this case each oxygen atom donates two electrons and the four electrons (two pairs) are shared between both atoms. Fig. 5.3 shows a simplified representation of an oxygen molecule. This is usually shown as O=O.

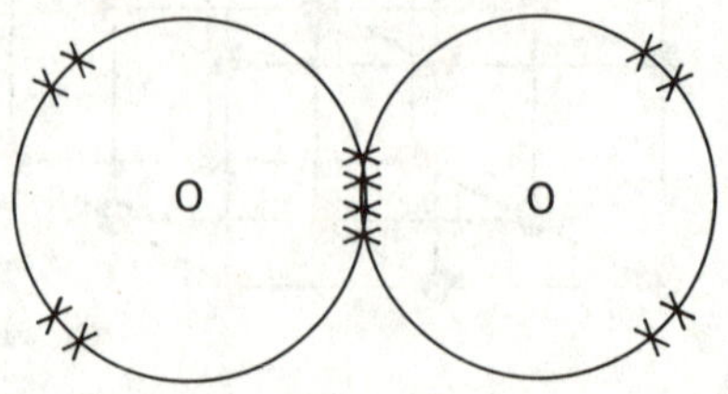

Fig. 5.3 Oxygen molecule

Other common examples of covalent bonding and the different molecular shapes are shown in Fig. 5.4.

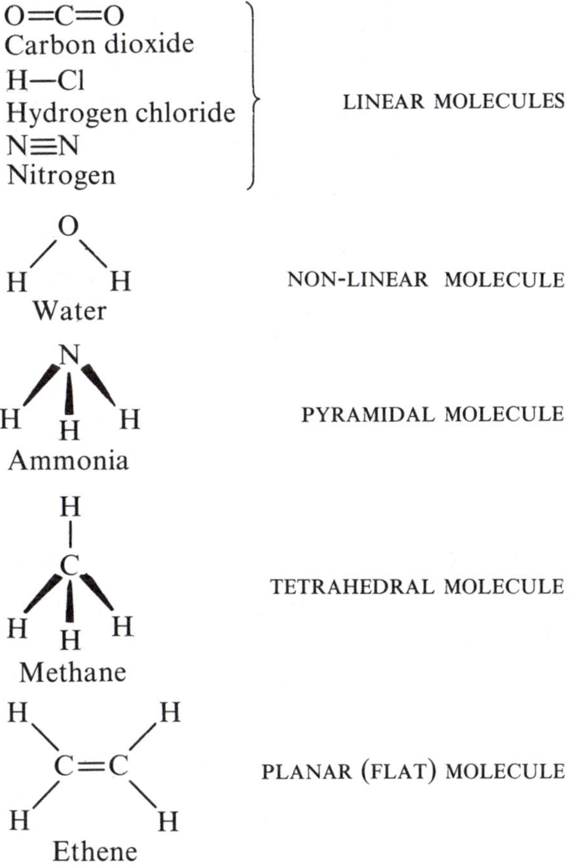

Fig. 5.4 Shapes of simple molecules

5.3 Co-ordinate (or dative) bonding

This is a much less common type of bonding but is closely related to covalent bonding. In co-ordinate bonding the bond formed between two atoms consists of **a pair of electrons both contributed by the same atom**, *e.g.* the formation of the ammonium ion.

H^+ and $:NH_3$ form a co-ordinate bond (represented by $H \leftarrow NH_3^+$) in which both electrons are donated by the nitrogen atom (Fig. 5.5). There are few examples of this type of bonding.

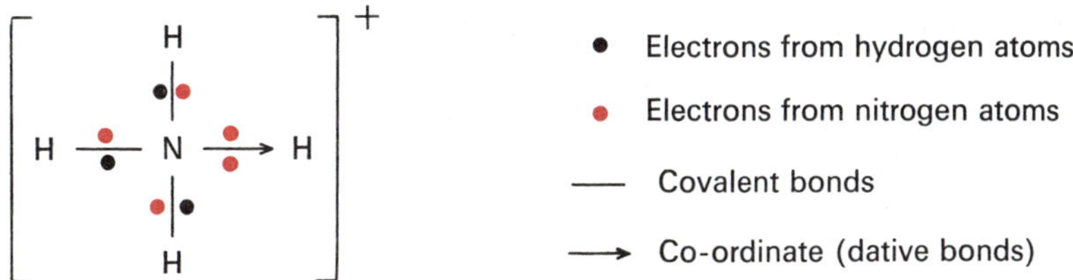

Fig. 5.5 Co-ordinate (dative) bonding in an ammonium ion

5.4 Metallic bonding

Metallic bonding is found only in metals. A metal consists of a close-packed regular arrangement of positive ions, which are surrounded by a 'sea' of electrons that bind the ions together. Fig. 5.6 shows the arrangement of ions in a single layer. There are two alternative ways of stacking these layers. The arrows in Fig. 5.6 indicate that the layer shown continues in all directions. Around any one ion in a layer there are six ions arranged hexagonally.

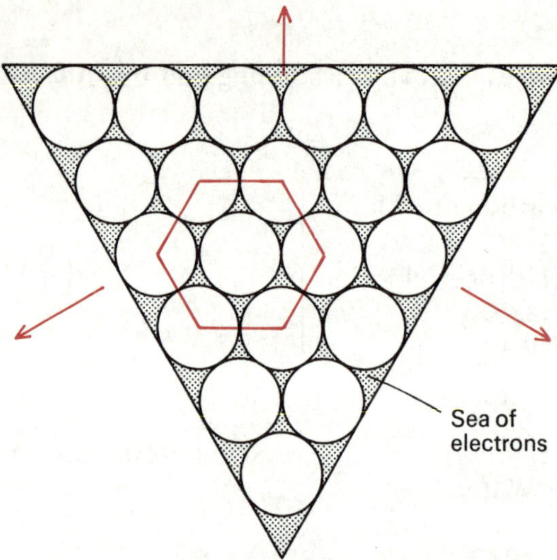

Fig. 5.6 Arrangement of ions in a metal layer

5.5 HYDROGEN BONDING

Hydrogen bonding is a weaker form of bonding, often between molecules. Water is a covalent molecule (Fig. 5.4). All chemical knowledge would lead us to expect water to be a gas. However, hydrogen bonds hold the water molecules together and so a higher temperature is needed before the water turns into a gas (Fig. 5.7).

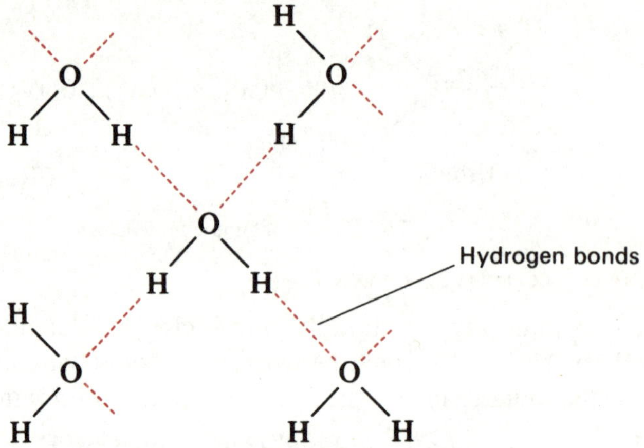

Fig. 5.7 Hydrogen bonding in water

Hydrogen bonding is caused by unequal sharing of the pair of electrons between each oxygen and hydrogen atom. This is caused by the greater **electronegativity** or 'electron pulling power' of the oxygen atom. It makes each oxygen atom slightly negatively charged and each hydrogen atom slightly positively charged. The hydrogen bonds are, therefore, weak electrostatic bonds. Hydrogen bonding is also found in alcohols.

5.6 EFFECTS OF BONDING ON PROPERTIES OF SUBSTANCES

Compounds containing ionic bonds have certain properties in common—they have high melting and boiling points, and as a result are solids at room temperature. The ions are tightly held together in a regular lattice and energy (called **lattice energy**) is required to break up the lattice and melt the substance. The melting point of magnesium oxide is very high making it suitable for use as a **refractory** (*i.e.* for lining furnaces).

Substances containing ionic bonding usually dissolve in water (a **polar solvent**) but not in a non-polar solvent. If they do not dissolve in water it is often because they have very high lattice energy.

Electricity passes through substances containing ionic bonds when the substances are molten or in solution in water but not when solid. These substances are called **electrolytes**.

Substances containing covalent bonding may be solid, liquid or gas at room temperature. They are usually insoluble in polar solvents but more soluble in non-polar solvents. They do not conduct electricity in any state.

Metals generally have high densities because the ions are close packed in the lattice. Because of the strong bonds between the ions caused by the **free electrons**, the melting points of most metals are high. The free electrons explain why metals are good conductors of heat and electricity.

Compounds containing hydrogen bonds have higher melting and boiling points than would otherwise be expected.

5.7 STRUCTURES OF SUBSTANCES

The different types of structure found in pure materials are summarised in Fig. 5.8.

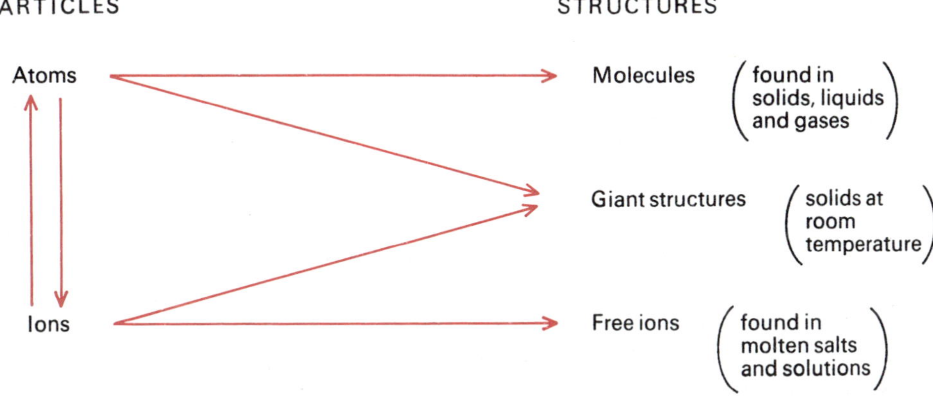

Fig. 5.8 Summary of structures

A substance with a **molecular structure** contains separate groups of atoms called **molecules**. The groups of atoms are tightly held together within the molecules by covalent bonds but the forces between the molecules are much weaker. If the molecules are small, the substance is usually a liquid or a gas. A few solids do contain small molecules, but these solids have low melting points, *e.g.* sulphur and iodine. If the molecules are large the substance will be a solid with a high melting point. Large molecules are sometimes called **macromolecules** (*e.g.* polythene) or **giant molecular structures** (*e.g.* silicon dioxide). Substances with a molecular structure do not conduct electricity in any state because there are no free ions or electrons. They are usually not very soluble in water but dissolve in non-polar solvents. Examples include sulphur, iodine, chlorine and carbon dioxide.

A substance with an ionic structure contains many ions bonded together into a **giant structure** of ions. Thus these substances always have high melting points as a large amount of energy is needed to break up the structure. Ionic substances conduct electricity when molten or dissolved in water because the ions are free to move. An example is sodium chloride.

Metals consist of **giant structures** (high melting points) except mercury (liquid). Carbon is a non-metal that has a giant atomic structure, and in the form of graphite it is also the only non-metal that conducts electricity. Most other non-metallic elements exist as small molecules, which is why many of them are gases under normal conditions. The noble gases exist as single atoms in a gaseous state.

5.8 ALLOTROPY

Allotropy is the existence of two or more forms of an element in the same physical state. These different forms are called **allotropes**. Allotropy is caused by the possibility of more than one arrangement of atoms. For example carbon can exist in two allotropes—diamond and graphite. Sulphur can exist in two allotropes—α-sulphur and β-sulphur.

6 States of matter

6.1 STATES OF MATTER

There are three states of matter—**solid, liquid** and **gas**. Any substance can exist in each of these three states depending on the conditions of temperature and pressure. Fig. 6.1 shows the relationship between these states of matter.

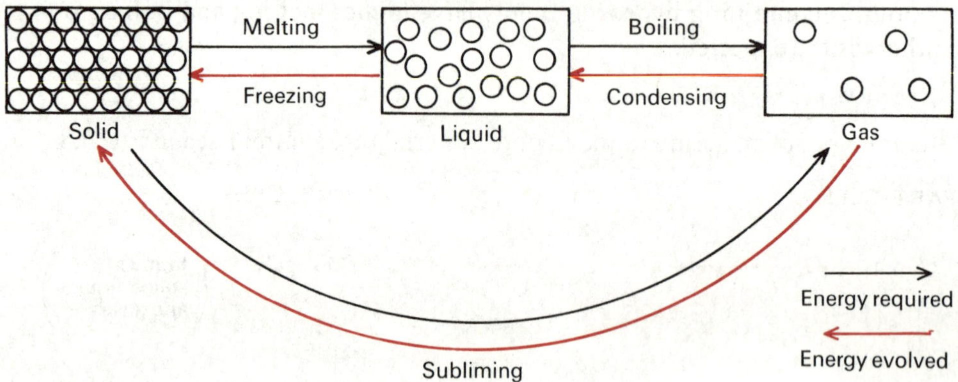

Fig. 6.1 States of matter and their interconversion

When a solid is heated it **melts** and forms a liquid. The temperature at which both solid and liquid can exist is called the **melting point** (or freezing point). When a liquid is heated to its **boiling point**, it **boils** and forms a gas (or vapour).

A simple two-dimensional representation of the particles in a solid, liquid and gas is shown in Fig. 6.1.

In the solid state the particles are usually **regularly** arranged and **rigidly** held in position. The particles can only vibrate. The particles vibrate more as temperature is increased.

In the liquid state the particles are able to move much more than in the solid. Liquids are usually less dense than the corresponding solids because the particles are more widely spaced. Ice is, however, exceptional because it is less dense than water at 0°C and therefore floats on water. When a solute is dissolved in a liquid, the solute particles fill the spaces between the particles of the liquid. In a liquid there are still forces holding the particles together.

Table 6.1 Comparison of solids, liquids and gases

Property	*Solid*	*Liquid*	*Gas*
Volume	Definite	Definite	Variable—expands or contracts to fill container
Shape	Definite	Takes up shape of bottom of container	Takes up the shape of the whole container
Density	High	Medium	Low
Expansion when heated	Low	Medium	High
Effect of applied pressure	Very slight	Slight decrease in volume	Large decrease in volume
Movement of particles	Very slow	Medium	Fast moving

In the ideal gas state the particles are completely independent and are moving randomly in all directions. As temperature increases the particles move faster and therefore collide more often. Gases are very compressible because of the large spaces between particles and this also causes the density to be low.

The properties of solids, liquids and gases are summarised in Table 6.1.

Water can exist in these three states—ice (solid), water (liquid) and steam (gas). Fig. 6.2 shows the graph obtained when a sample of ice is heated with a steady source of energy.

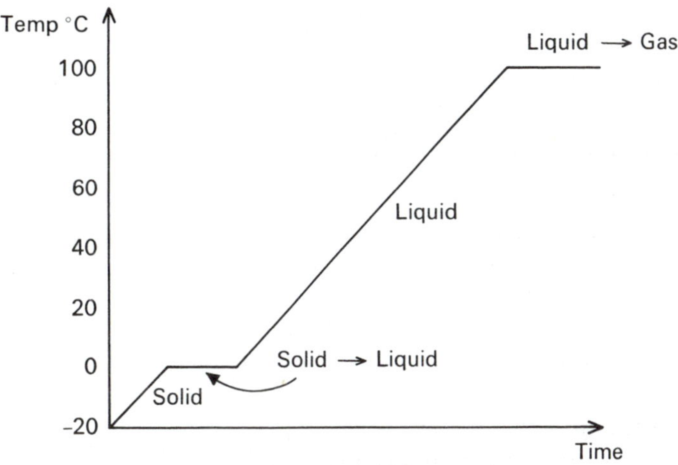

Fig. 6.2 Heating water

When a change of state is taking place, *e.g.* solid → liquid or liquid → gas, the temperature remains constant despite a continuing supply of energy. This energy, which is not being used to raise the temperature, is called **latent heat**. Latent heat is used to supply the particles with the extra energy they require as the state changes. It is evolved when the reverse changes take place, *e.g.* when steam condenses to form water.

6.2 Sublimation

Certain substances (*e.g.* ammonium chloride and solid carbon dioxide) do not melt when heated but change directly from a solid to a gas. As the gas cools it returns directly to the solid state. The reformed solid is chemically the same as the substance heated and is called the **sublimate**. This process is called **sublimation**, and it is a useful method for separating a mixture of two substances where only one substance sublimes, *e.g.* ammonium chloride and sodium chloride. Iodine is frequently quoted as an example of a substance that sublimes. However, when iodine crystals are heated they are usually seen to melt.

6.3 Kinetic theory

The fact that all particles in a solid, liquid or gas are in a state of constant motion is called the kinetic theory. It explains two fundamental concepts—**diffusion** and **brownian motion**.

6.4 Diffusion

If a drop of liquid bromine is dropped into a gas jar containing air and the gas jar covered, the liquid bromine vaporises and, after a while, the bromine vapour has spread evenly throughout the gas jar. The bromine particles and the constituent particles of the air have mixed thoroughly together. This spontaneous movement of particles to spread out to fill the whole container is called **diffusion**.

Diffusion also takes place in liquids and solids but it is much slower because the particles are moving more slowly in liquids and solids. If a crystal of purple potassium permanganate is dropped into water, the colour spreads throughout the water after about a week.

If a piece of cotton wool soaked in concentrated hydrochloric acid (evolving hydrogen chloride fumes) and a piece of cotton wool soaked in concentrated ammonia solution (evolving ammonia fumes) are put in the opposite ends of a dry 100 cm long glass tube, a ring forms after about five minutes as shown in Fig. 6.3.

$$NH_3(g) + HCl(g) \rightleftharpoons NH_4Cl(s)$$
$$\text{ammonia} + \text{hydrogen chloride} \rightleftharpoons \text{ammonium chloride}$$

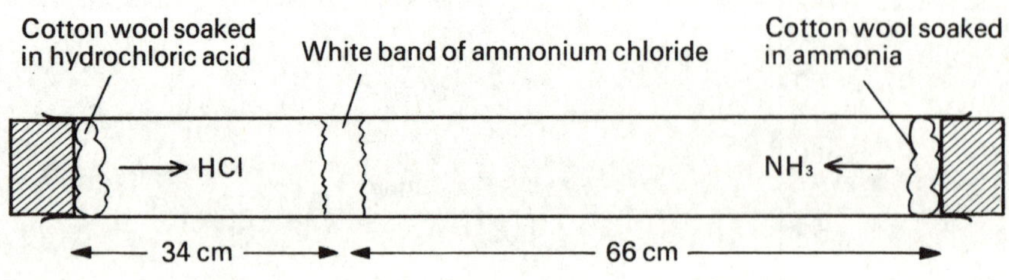

Fig. 6.3 Diffusion of ammonia and hydrogen chloride

The ring does not form immediately because:
(i) the particles are not moving just in one direction, and
(ii) the tube is filled with air.

The ammonia particles are moving about twice as fast as the hydrogen chloride particles and so the ring of ammonium chloride formed when the gases meet is nearer the piece of cotton wool soaked in concentrated hydrochloric acid.

Generally smaller (or lighter) particles move faster than larger (or heavier) particles.

6.5 Graham's Law of Diffusion

The rate of diffusion of a gas is inversely proportional to the square root of its density. Therefore if two gases diffuse under the same conditions of temperature and pressure, then

$$\frac{\text{rate of diffusion of gas 1}}{\text{rate of diffusion of gas 2}} = \frac{\sqrt{\text{density of gas 2}}}{\sqrt{\text{density of gas 1}}}$$

The density of a gas (*i.e.* the mass of 1 dm^3) is not the same as the vapour density. It is, however, proportional to the vapour density which is, in turn, proportional to the molecular mass. Thus

$$\frac{\text{rate of diffusion of gas 1}}{\text{rate of diffusion of gas 2}} = \frac{\sqrt{\text{mass of 1 mole of molecules of gas 2}}}{\sqrt{\text{mass of 1 mole of molecules of gas 1}}}$$

Example. In an experiment 100 cm^3 of methane (CH_4—mass of 1 mole of molecules = 16 g) diffuses in 24 seconds. Under the same conditions of temperature and pressure, 200 cm^3 of sulphur dioxide diffuses in 96 seconds. What is the mass of 1 mole of molecules of sulphur dioxide?

$$\text{Rate of diffusion of methane} = \frac{100}{24}\ cm^3/s$$

$$\text{Rate of diffusion of sulphur dioxide} = \frac{200}{96}\ cm^3/s$$

Substituting in the equation above

$$\frac{100/24}{200/96}=\frac{\sqrt{M}}{\sqrt{16}}$$ where M is the mass of 1 mole of molecules of sulphur dioxide

$$\sqrt{M}=4\times\frac{100}{24}\times\frac{96}{200}$$

$$\sqrt{M}=8$$

$$\therefore\ M=64$$

Mass of 1 mole of sulphur dioxide molecules = 64 g.

6.6 Brownian motion

In 1827, **Robert Brown** observed that fine pollen grains on the surface of water were not stationary but were in a state of constant random motion. The movement of these pollen grains is caused by collisions between water particles (these are too small to be seen) and pollen grains. Three facts are worth noting:

(*i*) The water particles must be moving rapidly in order to move the larger pollen grains.
(*ii*) The direction of movement of a pollen grain is determined by the frequency and direction of these collisions.
(*iii*) There is no pattern in the movement of pollen grains or water particles.

7 Some laws relating to gases

7.1 Introduction

Because volumes of gases are more susceptible to changes of temperature and pressure than liquids or solids, much study of the behaviour of gases has lead to several important gas laws.

7.2 Boyle's law

The volume of a fixed mass of gas at constant temperature is inversely proportional to the pressure.

i.e. as the pressure (P) increases the volume (V) decreases.
This is expressed mathematically as

$$P=\text{constant}\times\frac{1}{V}$$

7.3 Charles' law

The volume of a fixed mass of gas at constant pressure is directly proportional to the absolute temperature.

i.e. as the temperature (T) increases and the particles move faster, they move farther apart and the volume increases.

Mathematically $P=\text{constant}\times T$, where T is in degrees Kelvin.

(The temperature in degrees Kelvin (K) is found by adding 273° to the temperature Celsius

$$\therefore\ 20°\text{C}=20+273=293\text{K}.)$$

7.4 General gas equation

These two equations are combined in the general gas equation

$$\frac{P_1V_1}{T_1}=\frac{P_2V_2}{T_2}$$

This is useful for correcting the volume of a fixed mass of gas under one set of conditions of temperature and pressure to the volume of the same fixed mass of gas under a different set of conditions.

If a gas is at 0°C (273K) and 760 mm pressure, these conditions are called **standard temperature and pressure** (stp).

Example. A fixed mass of gas has a volume of 76 cm³ at 27°C and 750 mm pressure. Find the volume that the gas would have at stp

V_1 = volume before correction = 76 cm³
P_1 = pressure before correction = 750 mm
T_1 = temperature before correction = 27 + 273 = 300K
V_2 = volume after correction – unknown
P_2 = pressure after correction = 760 mm
T_2 = temperature after correction = 273K

Substitute in $$\frac{P_1V_1}{T_1}=\frac{P_2V_2}{T_2}$$

$$\frac{76 \times 750}{300}=\frac{760 \times V_2}{273}$$

$$V_2=\frac{273}{4}=68.25\text{cm}^3$$

Since both the pressure has increased and the temperature decreased, it is expected that the answer must be less than 76 cm³.

8 Air and combustion

8.1 THE COMPOSITION OF AIR

The composition of air varies from place to place because air is a mixture of gases. The composition of air by volume is approximately:

Nitrogen	78%
Oxygen	21%
Argon	1%

with small amounts of carbon dioxide, water vapour and other gases. Air does not usually contain hydrogen.

8.2 SEPARATING AIR INTO ITS CONSTITUENT GASES

Air is treated to remove carbon dioxide, water vapour and dust. The gas is cooled and compressed and it is then allowed to expand rapidly through a small nozzle. This produces a cooling effect on the air and, if repeated a number of times, leads to the air liquefying and forming liquid air.

When liquid air warms up the different components boil off at different temperatures. Nitrogen boils at –196°C and oxygen later at –183°C. This method of separation is called **fractional distillation of liquid air**, and it is possible because air is a **mixture** of gases.

8.3 PERCENTAGE OF OXYGEN IN AIR BY VOLUME

Fig. 8.1 shows the apparatus that can be used to find the percentage of oxygen accurately. An approximate value can be obtained by burning a piece of phosphorus in a fixed volume of air trapped over water.

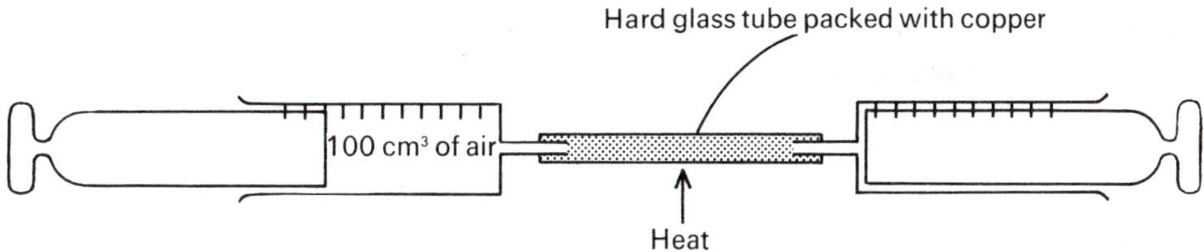

Fig. 8.1 Percentage of oxygen in air

100 cm^3 of air is trapped in one of the syringes. This air is passed backwards and forwards over heated copper turnings in a hard glass tube. The copper reacts with oxygen in the sample of air producing solid copper(II) oxide. The heating is continued until there is no further reduction in the volume of air. The apparatus is left to cool to room temperature and the volume of gas remaining (*i.e.* air minus the oxygen) measured.

E.g. Volume of gas before = 100 cm^3
Volume of gas after = 80 cm^3
∴ 100 cm^3 of air contains 20cm^3 of oxygen
Percentage of oxygen = 20%

8.4 PROCESSES INVOLVING GASES IN THE AIR

There are various processes that involve the gases in the air. These are summarised in Table 8.1.

Table 8.1 Processes that involve the gases in the air

Gas	*Combustion (see 8.5)*	*Rusting (see 8.6)*	*Respiration (see 8.7)*	*Photosynthesis (see 8.8)*
Nitrogen	Usually not involved	Not involved	Not involved	Not involved
Oxygen	Usually necessary	Necessary	Necessary	Produced
Carbon dioxide	Formed when carbon and carbon compounds burn	Speeds up rusting but it is not essential	Produced	Necessary
Noble (inert) gases	Not involved	Not involved	Not involved	Not involved
Water vapour	Formed when hydrogen and hydrogen compounds burn	Necessary	Produced	Necessary

8.5 COMBUSTION OR BURNING

The combustion or burning of a substance is the combination of the substance with oxygen. During the combustion, heat and light are usually given out. Substances burn better in pure oxygen than in air.

E.g. sulphur burns in air or oxygen with a blue flame

$$S(s) + O_2(g) \rightarrow SO_2(g)$$
sulphur + oxygen → sulphur dioxide

E.g. methane burns in excess air or oxygen

$$CH_4(g) + 2O_2(g) \rightarrow CO_2(g) + 2H_2O(g)$$
methane + oxygen → carbon dioxide + water

and in a limited supply of air or oxygen.

$$2CH_4(g) + 3O_2 \rightarrow 2CO(g) + 4H_2O(g)$$
methane + oxygen → carbon monoxide + water

Providing all the products are weighed and are not allowed to escape, a substance increases in mass during combustion.

8.6 RUSTING OF IRON

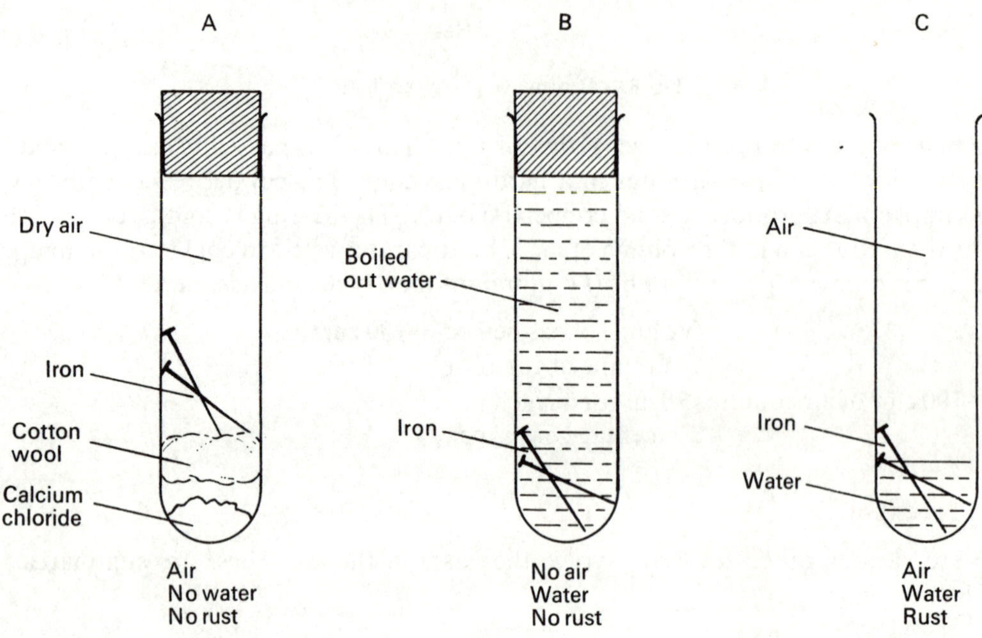

Fig. 8.2 Rusting of iron

If three test tubes are set up as shown in Fig. 8.2 and left for a few days, rusting only takes place in test tube C. This is because air and water are necessary for rusting. Rusting can be prevented by excluding air and/or water. Methods include greasing, painting and galvanising (coating the iron with a layer of zinc). Rust can be regarded as hydrated iron(III) oxide.

8.7 RESPIRATION OR BREATHING

Oxygen is transported from the lungs to the muscles by the haemoglobin in the blood. The oxygen is used to oxidise carbohydrates to form carbon dioxide.

E.g. $C_6H_{12}O_6(s) + 6O_2(g) \rightarrow 6CO_2(g) + 6H_2O(l) + \text{energy}$
glucose + oxygen → carbon dioxide + water + energy

The carbon dioxide is returned to the lungs and expelled from the body. Carbon monoxide is poisonous because it destroys the oxygen-carrying capacity of the haemoglobin (see 33.5).

8.8 PHOTOSYNTHESIS

Green plants absorb carbon dioxide from the air through the underside of their leaves. The carbon dioxide combines with water in the presence of sunlight and certain enzymes (see 20.10) to produce carbohydrates. The oxygen produced escapes into the air.

$$6CO_2(g) + 6H_2O(l) + \text{energy} \rightarrow C_6H_{12}O_6(aq) + 6O_2(g)$$
carbon dioxide + water + energy → glucose + oxygen

(N.B. This is the reverse of the equation in 8.7.)

8.9 CARBON CYCLE

The carbon cycle is shown in Fig. 8.3. Photosynthesis by plants ensures that the percentage of gases in the air remains fairly constant (carbon dioxide is used up while oxygen is produced—this is the opposite of respiration and the burning of fuels).

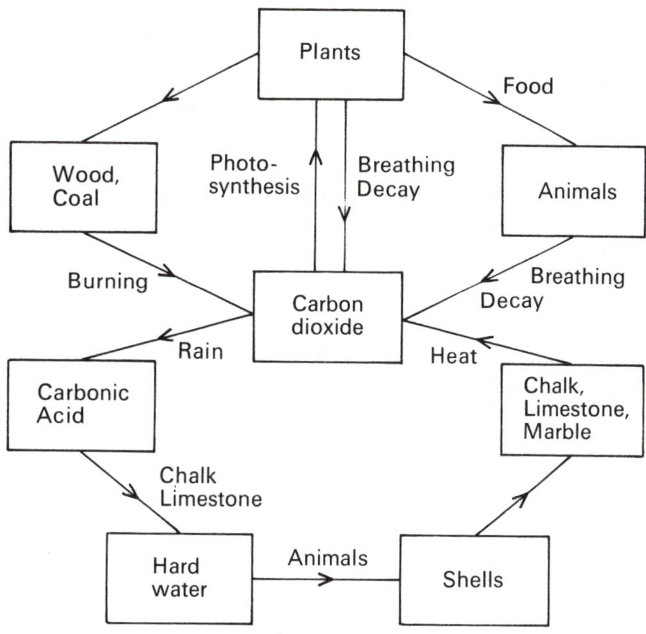

Fig. 8.3 The carbon cycle

---

9 Oxygen and oxides

9.1 Laboratory preparation of oxygen

Oxygen can be prepared in the laboratory by the decomposition of hydrogen peroxide solution using manganese(IV) oxide (manganese dioxide) as a catalyst.

$$2H_2O_2(aq) \rightarrow 2H_2O(l) + O_2(g)$$
$$\text{hydrogen peroxide} \rightarrow \text{water} + \text{oxygen}$$

The oxygen produced can be collected over water (Fig. 9.1).

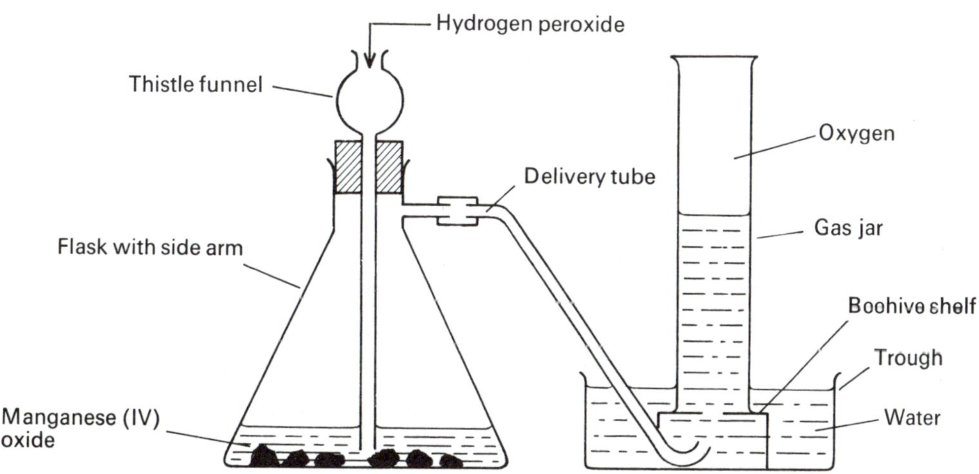

Fig. 9.1 Preparation of oxygen

Oxygen is produced industrially by the fractional distillation of liquid air (see 8.2).

9.2 PROPERTIES OF OXYGEN

Oxygen is a colourless, odourless and tasteless gas. It is neutral and only very slightly soluble in water. Its density is almost the same as the density of air (it cannot, therefore, be collected by upward or downward delivery).

All elements (apart from noble gases and some unreactive metals such as gold or platinum) combine with oxygen to form compounds called oxides. These reactions with oxygen are examples of oxidation.

Examples

(i) Magnesium. Magnesium ribbon burns brightly in oxygen to form a white solid, magnesium oxide. The oxide residue is alkaline when tested with universal indicator.

$$2Mg(s) + O_2(g) \rightarrow 2MgO(s)$$
$$\text{magnesium} + \text{oxygen} \rightarrow \text{magnesium oxide}$$

When magnesium burns in air a mixture of magnesium oxide and magnesium nitride is produced.

$$3Mg(s) + N_2(g) \rightarrow Mg_3N_2(s)$$
$$\text{magnesium} + \text{nitrogen} \rightarrow \text{magnesium nitride}$$

(ii) Sulphur. When sulphur is heated in air or oxygen it burns with a blue flame producing sulphur dioxide. Sulphur dioxide is acidic when tested with universal indicator.

$$S(s) + O_2(g) \rightarrow SO_2(g)$$
$$\text{sulphur} + \text{oxygen} \rightarrow \text{sulphur dioxide}$$

(iii) Iron. When iron wool is strongly heated in air or oxygen, it burns forming a black residue of iron(II) di-iron(III) oxide.

$$3Fe(s) + 2O_2(g) \rightarrow Fe_3O_4(s)$$
$$\text{iron} + \text{oxygen} \rightarrow \text{iron(II) di-iron(III) oxide}$$

(iv) Carbon. Carbon burns in air or oxygen. In excess air or oxygen, carbon dioxide is formed but, in a limited supply of air or oxygen, carbon monoxide is formed.

$$C(s) + O_2(g) \rightarrow CO_2(g)$$
$$\text{carbon} + \text{oxygen} \rightarrow \text{carbon dioxide}$$

$$2C(s) + O_2(g) \rightarrow 2CO(g)$$
$$\text{carbon} + \text{oxygen} \rightarrow \text{carbon monoxide}$$

(v) Copper. When copper is heated in air or oxygen a black surface coating of copper(II) oxide is formed.

$$2Cu(s) + O_2(g) \rightarrow 2CuO(s)$$
$$\text{copper} + \text{oxygen} \rightarrow \text{copper(II) oxide}$$

In order to make a pure sample of copper(II) oxide from copper it is necessary to react copper with excess nitric acid to produce copper(II) nitrate. The copper(II) nitrate is decomposed by heating.

$$Cu(s) + 4HNO_3(l) \rightarrow Cu(NO_3)_2(aq) + 2H_2O(l) + 2NO_2(g)$$
$$\text{copper} + \text{conc. nitric acid} \rightarrow \text{copper(II) nitrate} + \text{water} + \text{nitrogen dioxide}$$
$$2Cu(NO_3)_2(s) + \rightarrow 2CuO(s) + 4NO_2(g) + O_2(g)$$
$$\text{copper(II) nitrate} \rightarrow \text{copper(II) oxide} + \text{nitrogen dioxide} + \text{oxygen}$$

Many compounds burn in air or oxygen to produce more than one product. (For examples see 23.3 and 33.1.)

All substances burn better in oxygen than in air. When a metal burns in air or oxygen, it forms a neutral or alkaline oxide. Acidic oxides are only formed when non-metals burn in oxygen. The burning of an element in air or oxygen, and then testing the pH of the residue, is used to classify an element as a metal or non-metal.

Oxides can be classified into seven groups as discussed below.

9.3 Acidic oxides

E.g. sulphur dioxide SO_2, carbon dioxide CO_2

Acidic oxides are oxides of non-metals. They dissolve in water to produce acids.

E.g.

$$CO_2(g) + H_2O(l) \rightleftharpoons H_2CO_3(aq)$$
carbon dioxide + water ⇌ carbonic acid

$$SO_2(g) + H_2O(l) \rightleftharpoons H_2SO_3(aq)$$
sulphur dioxide + water ⇌ sulphurous acid

(An acid anhydride is a non-metal oxide which dissolves in water to produce an acid. Carbon dioxide is the acid anhydride of carbonic acid.)

Acidic oxides neutralise an alkali to form a salt and water only.

$$CO_2(g) + 2NaOH(aq) \rightarrow Na_2CO_3(aq) + H_2O(l)$$
carbon dioxide + sodium hydroxide → sodium carbonate + water

9.4 Basic oxides

E.g. calcium oxide CaO, copper(II) oxide CuO

A basic oxide is an oxide of a metal. It reacts with an acid to form a **salt and water only**.

E.g.

$$CaO(s) + 2HCl(aq) \rightarrow CaCl_2(aq) + H_2O(l)$$
calcium oxide + hydrochloric acid → calcium chloride + water

$$CuO(s) + 2HCl(aq) \rightarrow CuCl_2(aq) + H_2O(l)$$
copper(II) oxide + hydrochloric acid → copper(II) chloride + water

Some basic oxides react with water to form alkalis.

$$CaO(s) + H_2O(l) \rightarrow Ca(OH)_2(s)$$
calcium oxide + water → calcium hydroxide

9.5 Neutral oxides

E.g. nitrogen monoxide NO (nitric oxide)

A neutral oxide does not react with either an acid or an alkali to form a salt.

9.6 Amphoteric oxides

E.g. zinc oxide ZnO, aluminium oxide Al_2O_3

A few metal oxides can behave as either acidic or basic oxides depending upon circumstances. The oxides are called amphoteric oxides.

E.g. Zinc oxide will react with an acid or an alkali to form a salt and water.

$$ZnO(s) + 2HCl(aq) \rightarrow ZnCl_2(aq) + H_2O(l)$$
zinc oxide + hydrochloric acid → zinc chloride + water

In this reaction zinc oxide is acting as a basic oxide in the presence of an acid.

$$ZnO(s) + 2NaOH(aq) \rightarrow Na_2ZnO_2(aq) + H_2O(l)$$
zinc oxide + sodium hydroxide → sodium zincate + water

In this reaction zinc oxide is acting as an acidic oxide in the presence of an alkali.

Similarly with aluminium oxide:

With an acid

$$Al_2O_3(s) + 6HCl(aq) \rightarrow 2AlCl_3(aq) + 3H_2O(l)$$
aluminium oxide + hydrochloric acid → aluminium chloride + water

With an alkali

$$Al_2O_3(s) + 2NaOH(aq) \rightarrow 2NaAlO_2(aq) + H_2O(l)$$
aluminium oxide + sodium hydroxide → sodium aluminate + water

9.7 HIGHER OXIDES

E.g. lead(IV) oxide (lead dioxide) PbO_2, manganese(IV) oxide (manganese dioxide) MnO_2

Higher oxides are oxides which contain more oxygen than expected. The common oxide of lead is lead(II) oxide PbO. Lead(IV) oxide PbO_2 contains extra oxygen. Higher oxides are usually oxidising agents and they often decompose to produce oxygen.

E.g. lead(IV) oxide oxidises concentrated hydrochloric acid to chlorine.

$$PbO_2(s) + 4HCl(aq) \rightarrow PbCl_2(s) + 2H_2O(l) + Cl_2(g)$$
lead(IV) oxide + hydrochloric acid → lead(II) chloride + water + chlorine

9.8 PEROXIDES

E.g. sodium peroxide Na_2O_2

These are a type of higher oxide. When acidified with cold, dilute sulphuric acid, hydrogen peroxide is produced.

$$Na_2O_2(s) + H_2SO_4(aq) \rightarrow Na_2SO_4(aq) + H_2O_2(l)$$
sodium peroxide + sulphuric acid → sodium sulphate + hydrogen peroxide

Peroxides contain O_2^{2-} ions and are oxidising agents. Peroxides are formed only by reactive metals.

9.9 MIXED OXIDES

E.g. Pb_3O_4 red lead or dilead(II) lead (IV) oxide

These oxides behave as if they are mixtures of simpler oxides. Red lead can be considered to behave in the same way as a mixture of two parts of lead(II) oxide and one part of lead(IV) oxide (PbO_2).

9.10 SUBSTANCES PRODUCING OXYGEN ON HEATING

A number of substances decompose on heating to produce oxygen. These include:

lead(IV) oxide

$$2PbO_2(s) \rightarrow 2PbO(s) + O_2(g)$$
lead(IV) oxide → lead(II) oxide + oxygen

red lead

$$2Pb_3O_4(s) \rightarrow 6PbO(s) + O_2(g)$$
red lead → lead(II) oxide + oxygen

potassium permanganate

$$2KMnO_4(s) \rightarrow K_2MnO_4(s) + MnO_2(s) + O_2(g)$$
potassium permanganate → potassium manganate + manganese(IV) oxide + oxygen

potassium nitrate or sodium nitrate

$$2KNO_3(s) \rightarrow 2KNO_2(s) + O_2(g)$$
potassium nitrate → potassium nitrite + oxygen

$$2NaNO_3(s) \rightarrow 2NaNO_2(s) + O_2(g)$$
sodium nitrate → sodium nitrite + oxygen

mercury oxide

$$2HgO(s) \rightarrow 2Hg(l) + O_2(g)$$
mercury(II) oxide → mercury + oxygen

silver oxide

$$2Ag_2O(s) \rightarrow 4Ag(s) + O_2(g)$$
silver oxide → silver + oxygen

9.11 TEST FOR OXYGEN

If a glowing splint is put into a sample of oxygen, the splint re-lights. Dinitrogen monoxide (nitrous oxide N_2O) also re-lights a glowing splint. This is because the heat from the glowing splint decomposes dinitrogen monoxide.

$$2N_2O(g) \rightarrow 2N_2(g) + O_2(g)$$
dinitrogen monoxide → nitrogen + oxygen

However, dinitrogen monoxide is rarely met in the laboratory. It is also much more soluble in cold water than oxygen.

9.12 USES OF OXYGEN

Oxygen has a large number of uses including:

(*i*) In steel making. Oxygen is blown through molten iron to oxidise all impurities. Substances are then added in calculated quantities to produce steel of the correct specifications.
(*ii*) In hospitals, for divers etc., oxygen is used for breathing purposes.
(*iii*) For use with acetylene in oxy-acetylene cutting and welding equipment (see 33.18).

10 Hydrogen

10.1 LABORATORY PREPARATION OF HYDROGEN

Hydrogen is prepared by the action of dilute sulphuric acid on zinc.

$$Zn(s) + H_2SO_4(aq) \rightarrow ZnSO_4(aq) + H_2(g)$$
zinc + sulphuric acid → zinc sulphate + hydrogen

The apparatus is shown in Fig. 10.1.

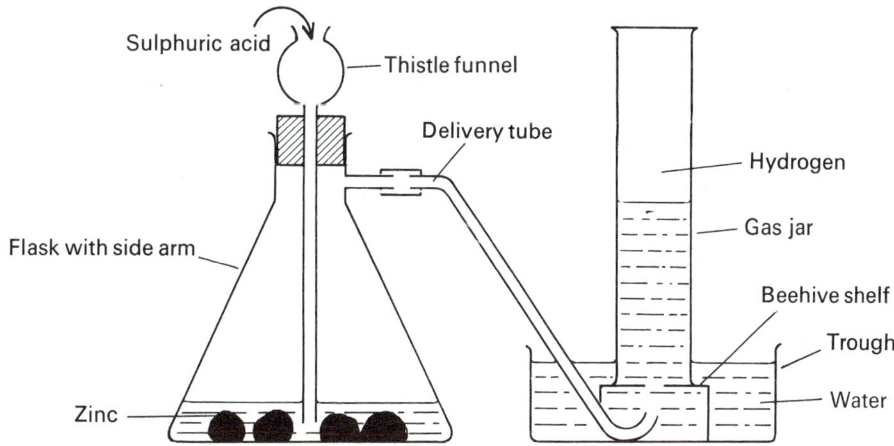

Fig 10.1 Preparation of hydrogen

Pure zinc reacts very slowly. The reaction can be speeded up by adding copper as a catalyst. (In practice, copper(II) sulphate solution is added.)

If the gas is required dry, it can be dried by passing it through concentrated sulphuric acid and collected by upward delivery.

10.2 TEST FOR HYDROGEN

When a lighted splint is put into hydrogen (preferably mixed with some air), it burns with a squeaky pop.

$$2H_2(g) + O_2(g) \rightarrow 2H_2O(g)$$
hydrogen + oxygen → water

10.3 OTHER LABORATORY REACTIONS PRODUCING HYDROGEN

(*i*) Reactions of metals with water (see 11.9).
(*ii*) Reactions of certain metals with alkalis.

Zinc, aluminium and silicon react with sodium hydroxide solution producing hydrogen.

$$Zn(s) + 2NaOH(aq) \rightarrow Na_2ZnO_2(aq) + H_2(g)$$
zinc + sodium hydroxide → sodium zincate + hydrogen

$$2Al(s) + 2NaOH(aq) + 2H_2O(l) \rightarrow 2NaAlO_2(aq) + 3H_2(g)$$
aluminium + sodium hydroxide + water → sodium aluminate + hydrogen

10.4 MANUFACTURE OF HYDROGEN

(*i*) Hydrogen is manufactured from coke. Steam is passed over heated coke producing **water gas** (a mixture of carbon monoxide and hydrogen). The water gas is mixed with more steam and passed over an iron-chromium catalyst at 500°C.

$$H_2(g) + CO(g) + H_2O(g) \rightarrow 2H_2(g) + CO_2(g)$$
water gas + water → hydrogen + carbon dioxide

The carbon dioxide is removed by dissolving it in water under pressure.

(*ii*) Hydrogen is manufactured from natural gas. Natural gas is mainly methane (CH_4). Methane is converted to hydrogen by partial oxidation at a high temperature (1300°C).

$$2CH_4(g) + O_2(g) \rightarrow 2CO(g) + 4H_2(g)$$
methane + oxygen → carbon monoxide + hydrogen

or by oxidation with steam in the presence of a nickel catalyst at 600°C.

$$CH_4(g) + H_2O(g) \rightarrow CO(g) + 3H_2(g)$$
methane + water → carbon monoxide + hydrogen

The mixture of carbon monoxide produced by either method is then treated as in (*i*).

10.5 PROPERTIES OF HYDROGEN

Hydrogen is a colourless, odourless, neutral gas. It has the lowest density of all the elements.

(i) Combustion of hydrogen in air
A jet of hydrogen burns in air to form steam which condenses to produce water.

$$2H_2(g) + O_2(g) \rightarrow 2H_2O(l)$$
hydrogen + oxygen → water

(ii) Reducing agent properties of hydrogen
If hydrogen is passed over heated copper(II) oxide, lead(II) oxide or iron(III) oxide (using the apparatus in Fig. 13.1), the metal oxides are reduced to the metals.

E.g.

$$CuO(s) + H_2(g) \rightarrow Cu(s) + H_2O(g)$$
copper(II) oxide + hydrogen → copper + water

$$PbO(s) + H_2(g) \rightarrow Pb(s) + H_2O(g)$$
lead(II) oxide + hydrogen → lead + water

$$Fe_2O_3(s) + 3H_2(g) \rightarrow 2Fe(s) + 3H_2O(g)$$
iron(III) oxide + hydrogen → iron + water

10.6 USES OF HYDROGEN

(*i*) Hydrogen is used in the manufacture of ammonia (see 27.3) and hydrochloric acid (see 32.5).
(*ii*) Hydrogen is used in the manufacture of margarine (see 33.15).
(*iii*) Hydrogen is used for filling balloons. It is, however, inflammable.

11 Water

11.1 Introduction

Water is a compound of hydrogen and oxygen (hydrogen oxide H_2O). It is a colourless, odourless liquid at room temperature and pressure.

The **water cycle** explains the regular supply of rain which provides the water essential for life. Evaporation of water from rivers, lakes and the sea provides water vapour which is held in the atmosphere in clouds. When the clouds cool the water vapour condenses and falls as rain.

The water supply to our homes comes from unpolluted rivers, lakes or suitable underground sources. It is not pure but contains a range of dissolved substances depending on the rocks through which the water has passed. The water is filtered and treated with a small quantity of chlorine to kill bacteria.

11.2 Water as a solvent

Water dissolves a wide range of different substances. Water is said to be a good **solvent** and the substances dissolved are called **solutes**. Water is a **polar solvent**, *i.e.* it contains small positive and negative charges caused by the slight movement of electrons in the covalent bonds. Polar solvents dissolve compounds containing ionic bonds, *e.g.* sodium chloride. Non-polar solvents (*e.g.* tetrachloromethane CCl_4) are poor at dissolving ionic compounds but are good solvents for molecular compounds.

A solution which contains as much solute as can be dissolved at a particular temperature is called a **saturated solution**. If any more solute is added to a saturated solution, the extra solute remains undissolved. The **solubility** of a solute is the mass of solute (in grams) which dissolves in 100 g of solvent at a particular temperature to form a saturated solution.

Generally the solubility of ionic compounds in water (*e.g.* potassium nitrate) increases as the temperature rises but the solubility of gases in water decreases as temperature rises. A graph of solubility versus temperature for a solute is called a **solubility curve**. Fig. 11.1 shows the solubility curves for some common solutes.

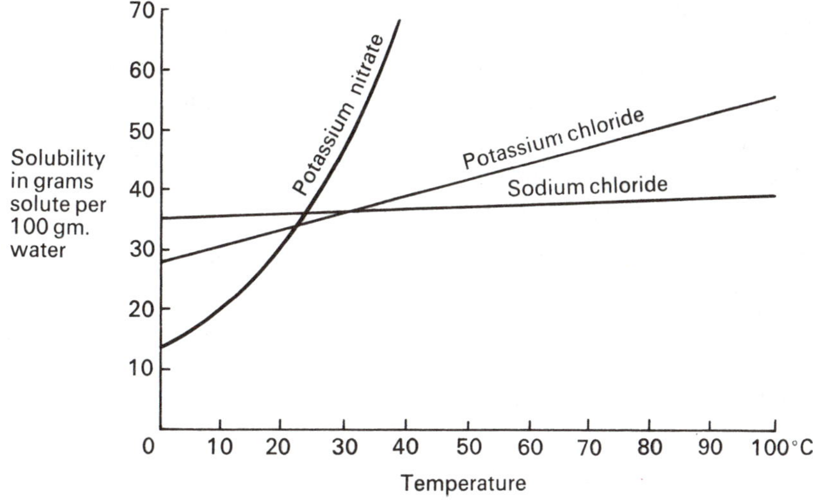

Fig. 11.1 Solubility curves

11.3 Hard and soft water

Distilled water (pure water) contains no dissolved solid impurities. It lathers well with soap. Rain water quite closely resembles distilled water.

Some water samples do not lather well with soap but form scum. These water samples are said to be hard. **Hard water** is caused by certain dissolved substances in water. These substances become dissolved as the water trickles through the ground. Water (containing dissolved carbon dioxide from the air) trickling through chalk (calcium carbonate) dissolves some of the chalk forming calcium hydrogencarbonate.

$$CaCO_3(s) + H_2O(l) + CO_2(g) \rightleftharpoons Ca(HCO_3)_2(aq)$$

calcium carbonate + water + carbon dioxide → calcium hydrogencarbonate

Hard water is caused by dissolved calcium and magnesium compounds in water.

There are two types of hardness in water—permanent hardness and temporary hardness. **Permanent hardness** is caused by dissolved **calcium sulphate** and **magnesium sulphate**. This type of hardness is not removed by boiling. It has to be softened (*i.e.* the hardness removed) by chemical reaction. **Temporary hardness** is caused by dissolved **calcium hydrogencarbonate**.

Apart from using more soap than would otherwise be required, using hard water has other effects. It causes deposits in kettles, boilers and hot water pipes. Hard water is, however, better for brewing beer and supplies calcium to the human body.

11.4 Removal of hardness

Temporary hardness is removed by boiling due to the decomposition of calcium hydrogencarbonate forming calcium carbonate, which is insoluble.

$$Ca(HCO_3)_2(aq) \rightleftharpoons CaCO_3(s) + H_2O(l) + CO_2(g)$$

calcium hydrogencarbonate ⇌ calcium carbonate + water + carbon dioxide

It is this deposit of calcium carbonate which forms the scale or 'fur' in a kettle.

Permanent and temporary hardness can be removed by adding washing soda crystals (sodium carbonate crystals).

$$Ca(HCO_3)_2(aq) + Na_2CO_3(aq) \rightarrow CaCO_3(s) + 2\,NaHCO_3(aq)$$

calcium hydrogencarbonate + sodium carbonate → calcium carbonate + sodium hydrogencarbonate

$$CaSO_4(aq) + Na_2CO_3(aq) \rightarrow CaCO_3(s) + Na_2SO_4(aq)$$

calcium sulphate + sodium carbonate → calcium carbonate + sodium sulphate

$$MgSO_4(aq) + Na_2CO_3(aq) \rightarrow MgCO_3(s) + Na_2SO_4(aq)$$

magnesium sulphate + sodium carbonate → magnesium carbonate + sodium sulphate

In each case the calcium or magnesium ions in solution are precipitated and no longer cause hardness problems. Other substances, *e.g.* calcium hydroxide, 'Calgon' (sodium metaphosphate) and sodium sesquicarbonate, work in a similar way by precipitating the substances which cause hardness. Sodium sesquicarbonate is better than sodium carbonate for household use because it is less alkaline.

Hardness can be removed by using an **ion-exchange column**. A column is filled with a suitable resin in small granules. The resin contains an excess of sodium ions. When the hard water passes through the column, the calcium and magnesium ions in the water (causing hardness) are exchanged for sodium ions. When the sodium ions on the column have all been removed, the column is recharged.

11.5 Soaps and soapless detergents

Soap is produced by treating vegetable or animal fats with concentrated sodium hydroxide solution, and precipitating the soap with salt solution. This reaction is an example of **saponification** (see 33.24).

Soapless detergents are produced from residues from crude oil distillation (see 33.20). These hydrocarbons are treated with concentrated sulphuric acid.

Soap and soapless detergent molecules are very similar in structure despite the different methods of production. Both contain a long hydrocarbon chain (*e.g.* $C_{17}H_{35}^-$) attached to an ionic group (CO_2^- in soap or SO_3^- in soapless detergents). The hydrocarbon 'tail'

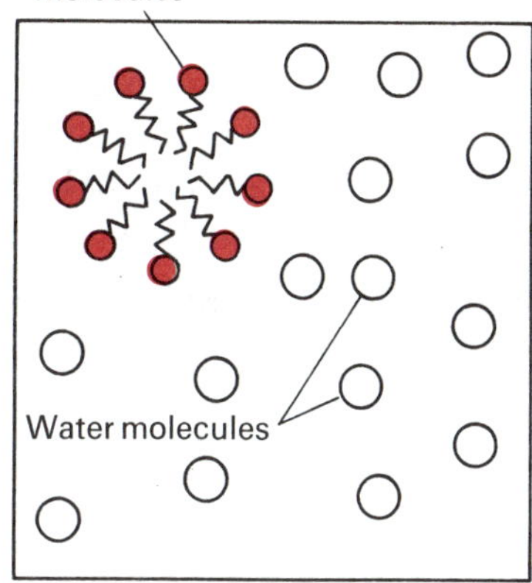

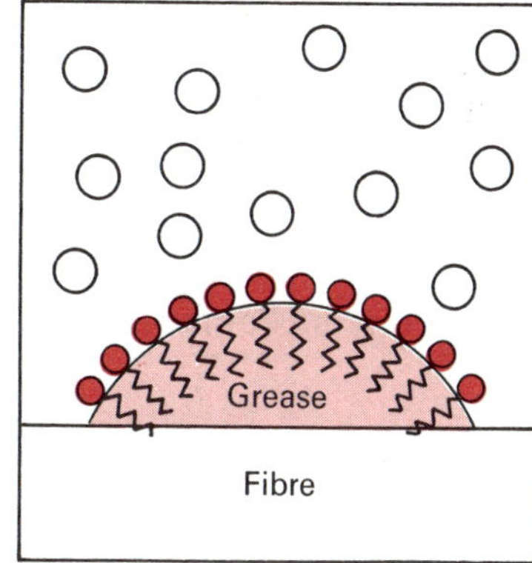

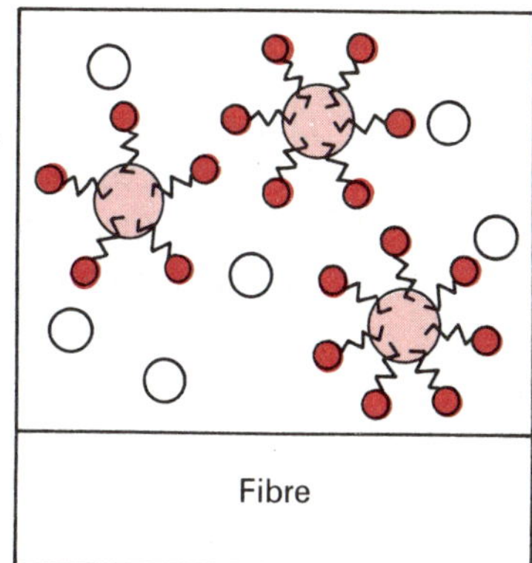

Fig. 11.2 Cleansing action of soap

will dissolve in fats and grease while the ionic 'head' will dissolve readily in water. The cleansing actions of soap or soapless detergents are summarised in Fig. 11.2. The soap or soapless detergent molecules are represented as 'tadpoles'. When the detergent (soap or soapless) is added to water the molecules are in clusters in the solution. The tails of the detergent molecules stick into the greasy dirt and attraction between the water molecules and the detergent molecules lifts the dirt from the fibre. Agitation of the solution helps to lift the dirt. The grease is then suspended in the solution, with repulsive forces between detergent molecules preventing grease from returning to the material.

Whereas soap forms scum with calcium and magnesium compounds in hard water, soapless detergents do not. This is an advantage of soapless detergents.

$$2C_{17}H_{35}CO_2^-Na^+(aq) + Ca^{2+}(aq) \rightarrow (C_{17}H_{35}CO_2)_2Ca(s) + 2\,Na^+(aq)$$

sodium stearate + calcium ions → calcium stearate + sodium ions
(soap) (scum)

11.6 Water of crystallisation

Certain substances crystallise (form crystals) with a fixed number of molecules of water contained within the crystal. This water is called **water of crystallisation**. For example, copper(II) sulphate crystallises as copper(II) sulphate crystals $CuSO_4.5H_2O$. They are blue in colour and are said to be **hydrated**. On heating, the water of crystallisation is evolved and white anhydrous copper(II) sulphate is formed. Other examples of substances containing water of crystallisation are:

hydrated sodium sulphate	$Na_2SO_4.10H_2O$
hydrated sodium carbonate	$Na_2CO_3.10H_2O$
hydrated magnesium sulphate	$MgSO_4.7H_2O$

There is no way of predicting the number of molecules of water of crystallisation in a particular substance.

11.7 Changes taking place when chemicals are exposed to the air

Some substances absorb water vapour from the air and are not appreciably changed. They are said to be **hygroscopic**. For example, copper(II) oxide absorbs small amounts of water vapour and should always be thoroughly dried before use in experiments involving weighing.

Other substances absorb water vapour to a much greater extent and the substance dissolves in the water absorbed. These substances are said to be **deliquescent**. An example of deliquescence is anhydrous calcium chloride. When left exposed to the atmosphere, the anhydrous calcium chloride absorbs water vapour from the air and after a couple of days a pool of calcium chloride solution remains.

Hygroscopic substances increase in mass by a small amount but deliquescent substances increase in mass considerably when left exposed to the atmosphere.

Some compounds containing water of crystallisation can lose some of this water on standing exposed to the air. This is called **efflorescence** (not to be confused with effervescence). Efflorescent compounds include hydrated sodium carbonate $Na_2CO_3.10H_2O$ and hydrated sodium sulphate $Na_2SO_4.10H_2O$. Hydrated sodium carbonate crystals are colourless, transparent crystals which on standing in the air turn to a white, opaque powder $Na_2CO_3.H_2O$. The mass of efflorescent compounds decreases on standing in air.

11.8 TESTING FOR WATER

Water is a colourless liquid but there are many other colourless liquids. When anhydrous copper(II) sulphate is added to pure water or any liquid containing water, it turns blue. Pure water freezes at 0°C and boils at 100°C (at atmospheric pressure).

11.9 REACTIONS OF METALS WITH WATER

The difference in reactivity of metals can be seen in the reactions of metals with water.

(i) Potassium

A small piece of potassium reacts violently with cold water. It floats on the water in a small molten ball and reacts with the water, producing hydrogen gas which catches alight and burns with a pinkish flame (see 41.3). An alkali (potassium hydroxide) remains in the solution.

$$2K(s) + 2H_2O(l) \rightarrow 2KOH(aq) + H_2(g)$$

potassium + water → potassium hydroxide + hydrogen

(ii) Sodium

Sodium reacts with cold water in a similar way to potassium but more slowly. The hydrogen does not usually ignite.

$$2Na(s) + 2H_2O(l) \rightarrow 2NaOH(aq) + H_2(g)$$

sodium + water → sodium hydroxide + hydrogen

(iii) Calcium

Calcium reacts slowly with cold water and usually sinks.

$$Ca(s) + 2H_2O(l) \rightarrow Ca(OH)_2(aq) + H_2(g)$$

calcium + water → calcium hydroxide + hydrogen

During the experiment the solution produced turns cloudy due to the formation of calcium hydroxide which is not very soluble and precipitates out.

(iv) Magnesium

Magnesium reacts very slowly with cold water. It takes several weeks to produce a test tube full of hydrogen from a short length of magnesium ribbon. It reacts well, however, when strongly heated in steam (Fig. 14.3).

$$Mg(s) + H_2O(g) \rightarrow MgO(s) + H_2(g)$$

magnesium + water(steam) → magnesium oxide + hydrogen

(N.B. Magnesium oxide is formed because, at the temperature of the experiment, magnesium hydroxide decomposes.)

(v) Zinc

Zinc reacts with steam in a similar way.

$$Zn(s) + H_2O(g) \rightarrow ZnO(s) + H_2(g)$$

zinc + water(steam) → zinc oxide + hydrogen

(vi) Iron

Iron, being less reactive, reacts only reversibly with steam.

$$3\,Fe(s) + 4\,H_2O(g) \rightleftharpoons Fe_3O_4(s) + 4\,H_2(g)$$
$$\text{iron} + \text{water (steam)} \rightleftharpoons \text{iron(II) di-iron(III) oxide} + \text{hydrogen}$$

Other metals lower in the reactivity series (*e.g.* lead, copper) do not react with water.

12 The mole and chemical calculations

This section is the basis of most chemical calculations that appear on examination papers. Questions based on this section appear frequently and provide a useful discrimination between candidates. In the Scottish O grade syllabus, the 'mole' concept is restricted to that of gram formula weight and the definition of the mole as the Avogadro number of particles is deferred until the Higher grade.

12.1 RELATIVE ATOMIC MASS

All atoms are too small to be weighed individually. It is possible, however, to compare the mass of one atom with the mass of another. This is done using a mass spectrometer. A magnesium atom has twice the mass of a carbon-12 atom and six times the mass of a helium atom.

The **relative atomic mass** of an atom is the number of times an atom is heavier than a hydrogen atom (or one twelfth of a carbon-12 atom). Relative atomic masses are not all whole numbers because of the existence of isotopes (see 4.1).

$$\text{Relative atomic mass} = \frac{\text{mass of 1 atom of element}}{\text{mass of 1 atom of hydrogen}}$$

The relative atomic mass is simply a number and has no units. You are not expected to remember relative atomic masses. They are given on examination papers in one of the following ways:

(*i*) (Ca = 40, C = 12)
or (*ii*) $A_r(Ca) = 40$

Relative atomic masses are sometimes just called atomic masses or atomic weights.

12.2 THE MOLE

As an alternative to comparing masses of individual atoms of different elements, it is possible to consider large numbers of atoms. For example:

1 atom of magnesium weighs twice as much as 1 atom of carbon-12
2 atoms of magnesium weigh twice as much as 2 atoms of carbon-12
100 atoms of magnesium weigh twice as much as 100 atoms of carbon-12

The mass of magnesium atoms will always be twice the mass of the carbon-12 atoms, providing the number of magnesium and carbon-12 atoms are the same.

We are used to collective terms to describe a number of objects, *e.g.* a dozen eggs, a gross of test tubes *etc.* In chemistry, the term **mole** is used in the same way. We speak of a mole of magnesium atoms, a mole of carbon dioxide or a mole of electrons.

A mole provides a quantity of material that can be used in the laboratory. A mole of carbon atoms (12 g) is just a small handful.

A mole contains approximately 6×10^{23} particles (600 000 000 000 000 000 000 000). This number is called **Avogadro's number (L)**. A very large number is consequently difficult for us to appreciate. If one were to stand on a sandy beach and look along the beach in both directions, you would not see enough particles of sand to make 1 mole of grains of sand.

It is conveniently arranged that 1 mole of atoms of any element has a mass equal to the relative atomic mass (but with units of g).

E.g. Relative atomic mass of Na = 23
∴ Mass of 1 mole of sodium atoms = 23 g

The mole may be defined as the amount of substance which contains as many elementary units as there are in 1 g of hydrogen (or 12 g of carbon-12). These elementary units can be considered as:

atoms	*e.g.* Mg, C, He
molecules	*e.g.* CH_4, H_2O
ions	*e.g.* Na^+, Cl^-
specified formula units	*e.g.* H_2SO_4

There are other terms that might be seen in books or examination papers:

E.g.
1 gram-atom—mass of 1 mole of atoms
1 gram-molecule—mass of 1 mole of molecules
1 gram-ion—mass of 1 mole of ions
1 gram-formula—mass of 1 mole of formula unit
1 Faraday—1 mole of electrons

The term '1 mole of chlorine' can be ambiguous. It could mean 1 mole of chlorine atoms (6×10^{23} atoms) or 1 mole of chlorine molecules (12×10^{23} atoms).

12.3 VOLUME OF ONE MOLE OF MOLECULES OF A GAS

There is no simple relationship that predicts the volume occupied by 1 mole of molecules in a solid or a liquid. However, 1 mole of molecules of any gas occupies 24 000 cm^3 (24 dm^3) at room temperature and pressure or 22 400 cm^3 (22.4 dm^3) at stp (see 7.4). This information is given on the examination paper if it is required.

12.4 MOLAR SOLUTIONS

When 1 mole of a substance is dissolved in water and the volume of solution made up to 1000 cm^3 (1 dm^3), the resulting solution is called a **molar (or M) solution**. If 2 moles of a substance are made up to 1000 cm^3 (or 1 mole made up to 500 cm^3) the solution is said to be a 2M solution. *E.g.* 8 g of sodium hydroxide NaOH is dissolved in water to make 100 cm^3 of solution. What is the molarity of the solution? (Na = 23, O = 16, H = 1)

Mass of 1 mole of sodium hydroxide NaOH = 23 + 16 + 1 = 40 g
8 g of sodium hydroxide is $\frac{8}{40}$ = 0.2 mole (see 12.5)

0.2 mole of sodium hydroxide dissolved to make 100 cm^3 of solution.
0.2×10 mole of sodium hydroxide dissolved to make 1000 cm^3 of solution.

Solution produced is then 0.2×10 M, *i.e.* 2M sodium hydroxide.

12.5 CONVERSION OF NUMBER OF GRAMS OF SUBSTANCE TO NUMBER OF MOLES

It is useful to convert, for example, 18 g of water to 1 mole of water molecules. This is because we know that there are 6×10^{23} water molecules in 1 mole of water but we have no idea how many particles there are in a given mass of water.

$$\text{No. of moles} = \frac{\text{Number of grams}}{\text{Mass of 1 mole}}$$

E.g. How many moles of carbon dioxide molecules are present in 11 g of carbon dioxide? (C = 12, 0 = 16)

$$\text{Mass of 1 mole of carbon dioxide } CO_2 = 12 + (2 \times 16)\text{ g}$$
$$= 44\text{ g}$$

$$\text{Number of moles of carbon dioxide} = \frac{11}{44}$$
$$= 0.25\text{ moles}$$

12.6 Conversion of number of moles to number of grams of substance

This is the reverse of 12.5.

Number of grams = Number of moles × Mass of 1 mole

E.g. What is the mass of 2 moles of ethanol molecules (C_2H_5OH)?

$$\text{Mass of 1 mole of ethanol molecules} = (2 \times 12) + (5 \times 1) + 16 + 1\text{ g}$$
$$= 46\text{ g}$$
$$\text{Number of grams of ethanol} = 2 \times 46\text{ g}$$
$$= 92\text{ g}$$

12.7 Finding the formula of a compound

In Unit 3 there is information to enable you to work out the formula of a compound. It should be remembered that each formula could be worked out following a suitable experiment involving weighing.

E.g. Magnesium oxide
A known mass of magnesium ribbon is burnt in a crucible in contact with air. The mass of magnesium oxide produced is found.

① Mass of crucible and lid	= 20.12 g	
② Mass of crucible, lid and magnesium	= 20.36 g	
Mass of magnesium	= 0.24 g	*i.e.* ② − ①
③ Mass of crucible, lid and magnesium oxide	= 20.52 g	
Mass of magnesium oxide	= 0.40 g	*i.e.* ③ − ①

0.24 g of magnesium combines with 0.16 g of oxygen (0.40 − 0.24) to form 0.40 g of magnesium oxide.

(This is the key statement that you must write down. It will help you but also will give you marks.)

Multiply through this statement by 100 to remove decimals and prevent arithmetical mistakes:

24 g of magnesium combines with 16 g of oxygen to form 40 g of magnesium oxide.

Divide masses of magnesium and oxygen by the appropriate relative atomic masses to give the number of moles of atoms of magnesium and oxygen (see 12.5) (Mg = 24, O = 16)

$\frac{24}{24}$ moles of magnesium atoms combine with $\frac{16}{16}$ moles of oxygen atoms.

1 mole of magnesium atoms combine with 1 mole of oxygen atoms. Since 1 mole of magnesium atoms contains the same number of atoms as 1 mole of oxygen atoms (*i.e.* 6×10^{23}), the simplest formula of magnesium is MgO. Since these questions are very common, here are two further examples:

1. 2.00 g of mercury combines with 0.71 g of chlorine to form 2.71 g of a mercury chloride. What is the simplest formula for the mercury chloride? (Hg = 200, Cl = 35.5)

Key statement:
2.00 g of mercury combines with 0.71 g of chlorine to form 2.71 g of mercury chloride.
Multiply throughout by 100 to remove decimals

200 g of mercury combines with 71 g of chlorine

Divide by the appropriate atomic masses.

$\frac{200}{200}$ moles of mercury atoms combine with $\frac{71}{35.5}$ moles of chlorine atoms.

1 mole of mercury atoms combines with 2 moles of chlorine atoms.

Simplest formula is $HgCl_2$.

2. 11.2 g of iron combines with 4.8 g of oxygen to form an iron oxide. What is the simplest formula for the iron oxide? (Fe = 56, O = 16)

Key statement:
11.2 g of iron combines with 4.8 g of oxygen to form 16.0 g of iron oxide.

Multiply throughout by 10 to remove decimals.

112 g of iron combines with 48 g of oxygen.

Divide by the appropriate relative atomic masses.

$\frac{112}{56}$ moles of iron atoms combine with $\frac{48}{16}$ moles of oxygen atoms.

2 moles of iron atoms combine with 3 moles of oxygen atoms.

Simplest formula is Fe_2O_3.

12.8 Calculating the simplest formula from percentages

E.g. A hydrocarbon contains 75% carbon and 25% hydrogen. Calculate the simplest and the molecular formulae for this compound given that the mass of 1 mole of molecules is 16 g. (C = 12, H = 1)

	C	H
Percentage	75%	25%
Relative atomic mass	12	1
Divide percentage by relative atomic mass	6.25	25
Divide by smallest, *i.e.* 6.25	1	4

Simplest formula (or empirical) formula = CH_4

This may not be the molecular formula. It could be C_2H_8, C_3H_{12} *etc.*—always 4 times as many hydrogens as carbons.

If the formula is CH_4, the mass of 1 mole of molecules is 16 g

Molecular formula is CH_4

12.9 Calculating the percentages of elements in a compound

E.g. Calculate the percentage of nitrogen in ammonium nitrate NH_4NO_3. (N = 14, H = 1, 0 = 16)

Mass of 1 mole of ammonium nitrate $NH_4NO_3 = 14 + (4 \times 1) + 14 + (3 \times 16)$
$= 80$ g

Each 80 g of ammonium nitrate contains 28 g of nitrogen (it contains 2 nitrogen atoms i.e. 2 × 14)

$$\text{Percentage of nitrogen} = \frac{28}{80} \times 100$$
$$= 35\%$$

This type of calculation is useful for calculating the percentage of nitrogen in a fertiliser.

12.10 Calculations from equations

This type of question appears frequently on O level and CSE papers. It is necessary to have a balanced equation. Usually this equation is supplied.

E.g. $Na_2CO_3(s) + 2HCl(aq) \rightarrow 2\,NaCl(aq) + H_2O(l) + CO_2(g)$
sodium carbonate + hydrochloric acid → sodium chloride + water + carbon dioxide

This equation gives the following information:

1 mole of sodium carbonate (Na_2CO_3) reacts with 2 moles of hydrochloric acid (HCl) to produce 2 moles of sodium chloride (NaCl), 1 mole of water (H_2O) and 1 mole of carbon dioxide (CO_2).

Using the relative atomic masses (Na = 23, C = 12, O = 16, H = 1, Cl = 35.5) we can find the masses of the substances that react together and the masses of the substances produced.

$$Na_2CO_3(s) + 2\,HCl(aq) \rightarrow 2\,NaCl(aq) + H_2O(l) + CO_2(g)$$

$(2 \times 23) + 12 + (3 \times 16)$	$2(1 + 35.5)$	$2(23 + 35.5)$	$(2 \times 1) + 16$	$12 + (2 \times 16)$
106 g	73 g	117 g	18 g	44 g

At this stage it is worthwhile checking that the sum of the masses on the left hand side (106 g + 73 g = 179 g) equals the right hand side, (117 g + 18 g + 44 g = 179 g). This is expected from the Law of Conservation of Mass (see 13.1). Silly arithmetical mistakes, frequently seen on examination papers, should be avoided if you do this.

The calculations now are just proportion sums.

1. Calculate the maximum mass of sodium chloride that could be produced from 5.3 g of sodium carbonate.

From the information following the equation:

106 g of sodium carbonate (1 mole) produces 117 g of sodium chloride (2 moles)

1 g of sodium carbonate produces $\frac{117}{106}$ g of sodium chloride

5.3 g of sodium carbonate produces $\frac{117 \times 5.3}{106}$ g of sodium chloride $= \frac{11.7}{2} = 5.85$ g

2. Calculate the volume of 2M hydrochloric acid which would exactly react with 5.3 g of sodium carbonate.

Using the information above again:

106 g of sodium carbonate (1 mole) reacts with 73 g of hydrochloric acid (2 moles).

Since hydrochloric acid is in a dilute solution it is worth remembering that:

1 mole of hydrochloric acid dissolved and made up to 1000 cm^3 (1 dm^3) produces a M solution (see 12.4).

2 moles of hydrochloric acid dissolved and made up to 1000 cm^3 produces a 2 M solution

∴ 106 g of sodium carbonate reacts with 1000 cm^3 of 2 M hydrochloric acid

1 g of sodium carbonate reacts with $\frac{1000}{106}$ cm^3 of 2 M hydrochloric acid

5.3 g of sodium carbonate reacts with $\frac{1000}{106} \times 5.3$ cm^3 of 2 M hydrochloric acid

= 50 cm^3 of 2 M hydrochloric acid.

3. Calculate the volume of carbon dioxide (at room temperature and pressure) produced when 5.3 g of sodium carbonate reacts with excess hydrochloric acid. Using the information above again:

106 g of sodium carbonate (1 mole) produces 44 g of carbon dioxide (1 mole). However, 1 mole of any gas at room temperature and pressure occupies 24 000 cm^3 (24 dm^3) (see 12.3).

∴ 106 g of sodium carbonate produces 24 000 cm^3 of carbon dioxide

1 g of sodium carbonate produces $\frac{24\,000}{106}$ cm^3 of carbon dioxide

5.3 g of sodium carbonate produces $\frac{24\,000 \times 5.3}{106}$ cm^3 of carbon dioxide

= 1200 cm^3

(All measurements at room temperature and pressure.)

The chemical calculations using equations enable a chemist to calculate the quantities of materials required for a particular reaction and to calculate the quantities of products formed.

13 Laws of chemical combination

This unit is written in two parts.

Part A 13.1–13.3 Laws relating to reacting masses
Part B 13.4–13.7 Laws relating to combining volumes of gases

13.1 Law of conservation of mass

There is no increase or decrease in mass during a chemical reaction. Matter cannot be created or destroyed. If the reacting substances are accurately weighed before reaction and the products accurately weighed after the reaction, the mass is unchanged.

For example, if equal volumes of sodium chloride and silver nitrate solutions are mixed, the following reaction takes place:

$$AgNO_3(aq) + NaCl(aq) \rightarrow AgCl(s) + NaNO_3(aq)$$
silver nitrate + sodium chloride → silver chloride + sodium nitrate

There is no mass change during the reaction.

The law applies providing no product escapes and the mass of all products is found. In the reaction between calcium carbonate and dilute hydrochloric acid, there is no change in mass providing the carbon dioxide does not escape.

$$CaCO_3(s) + 2\,HCl(aq) \rightarrow CaCl_2(aq) + H_2O(l) + CO_2(g)$$
calcium carbonate + hydrochloric acid → calcium chloride + water + carbon dioxide

If the reaction is carried out in a beaker, open to the atmosphere, there is a steady decrease in mass as the reaction proceeds due to the escape of carbon dioxide gas.

13.2 Law of constant composition

Some compounds can be prepared by a variety of methods. **The chemical composition of a compound is always the same** regardless of the method used.

Copper(II) oxide can be prepared by

(*i*) Action of heat on copper(II) nitrate cyrstals.

$$2Cu(NO_3)_2(s) \rightarrow 2CuO(s) + 4NO_2(g) + O_2(g)$$
copper(II) nitrate → copper(II) oxide + nitrogen dioxide + oxygen

(*ii*) Action of heat on copper(II) carbonate.

$$CuCO_3(s) \rightarrow CuO(s) + CO_2(g)$$
copper carbonate → copper(II) oxide + carbon dioxide

(*iii*) Precipitation of copper hydroxide followed by heating.

$$CuSO_4(aq) + 2NaOH(aq) \rightarrow Cu(OH)_2(s) + Na_2SO_4(aq)$$
copper(II) sulphate + sodium hydroxide → copper(II) hydroxide + sodium sulphate

$$Cu(OH)_2(s) \rightarrow CuO(s) + H_2O(g)$$
copper(II) hydroxide → copper(II) oxide + water

The copper(II) oxide produced by these three methods can be converted to copper by reduction with hydrogen (Fig. 13.1).

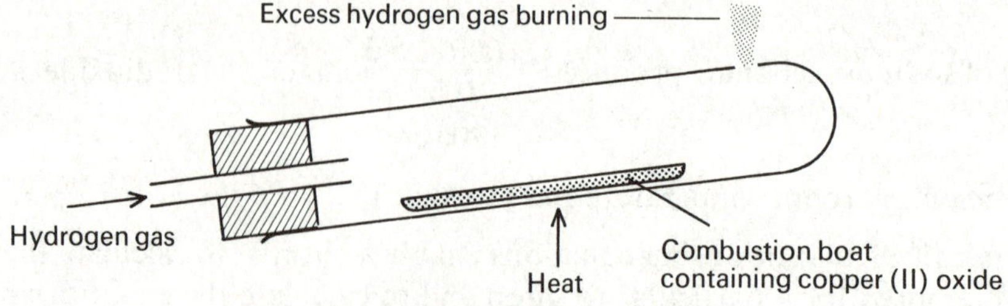

Fig. 13.1 Reduction of copper(II) oxide

A known mass of copper(II) oxide is heated in a stream of hydrogen gas until all the copper(II) oxide is reduced to copper.

$$CuO(s) + H_2(g) \rightarrow Cu(s) + H_2O(g)$$
copper(II) oxide + hydrogen → copper + water

Equal masses of copper(II) oxide, produced by different methods, when reduced to copper, form equal masses of copper. This law refers to the formation of just one compound (*e.g.* copper(II) oxide).

13.3 LAW OF MULTIPLE PROPORTIONS

The law of multiple proportions applies when two elements form more than one compound, *e.g.* copper(II) oxide CuO and copper(I) oxide Cu_2O.

If two elements (A and B) combine together to form more than one compound, then the different masses of A, that combine with a fixed mass of B, are in a simple ratio.

If known masses of the two copper oxides are reduced by hydrogen (using the apparatus in Fig. 13.1) to copper, the masses of copper that combine with 1 g of oxygen are

copper(I) oxide 8 g
copper(II) oxide 4 g

These two masses are in a simple ratio of 2:1.

Another frequently quoted example of this law is lead oxide. There are three lead oxides, PbO, PbO_2, Pb_3O_4. The masses of lead combining with a fixed mass of oxygen are in a ratio 4:2:3.

13.4 AVOGADRO'S LAW

This law applies to gases. **Equal volumes of different gases, under the same conditions of temperature and pressure, contain the same number of molecules.** This law is frequently used in conjunction with Gay Lussac's Law (see 13.5).

13.5 GAY LUSSAC'S LAW

The volumes of gases reacting and the volumes of gaseous products formed bear a simple numerical relation to one another, providing all measurements are made at the same temperature and pressure.

For example, by experiment it can be found that 10 cm^3 of nitrogen gas (1 volume) combines with 30 cm^3 of hydrogen gas (3 volumes) to produce 20 cm^3 of ammonia gas (2 volumes).

A frequent mistake at this level is to assume that the reaction of 10 cm^3 of nitrogen and 30 cm^3 of hydrogen must produce 40 cm^3 of ammonia. Although there is no change in the overall mass during the reaction (see 13.1), there is no necessity for the volume to remain constant.

E.g. Find the volume of oxygen required and the volume of gaseous products formed when 10 cm^3 of methane CH_4 is burnt in sufficient oxygen for complete combustion. All measurements were made at room temperature and pressure. First write the equation:

$$CH_4(g) + 2O_2(g) \rightarrow CO_2(g) + 2H_2O(l)$$
methane + oxygen → carbon dioxide + water
1 vol 2 vol → 1 vol 0

∴ When 10 cm^3 of methane gas (CH_4) reacts, 20 cm^3 of oxygen is required.
Volume of carbon dioxide produced = 10 cm^3.

N.B. When gases are cooled to room temperature, the steam condenses and the volume of water produced is negligible.

13.6 To find the formula of hydrogen chloride

It can be found by experiment that:

1 volume of hydrogen reacts with 1 volume of chlorine to produce 2 volumes of hydrogen chloride.

If 1 volume of hydrogen contains x molecules of hydrogen,
1 volume of chlorine contains x molecules of chlorine and
2 volumes of hydrogen chloride contain 2x molecules of hydrogen chloride.
(This uses Avogadro's Law.)

$\therefore$ 1 molecule of hydrogen combines with 1 molecule of chlorine to form 2 molecules of hydrogen chloride.

If we assume that molecules of hydrogen and chlorine each contain two atoms (*i.e.* they are diatomic), then 2 atoms of hydrogen combine with 2 atoms of chlorine to form 2 molecules of hydrogen chloride.

The formula of hydrogen chloride is HCl.

13.7 To find the formula of ammonia

When 40 cm^3 of ammonia gas are decomposed by passing over red hot iron wool, 60 cm^3 of hydrogen and 20 cm^3 of nitrogen are produced (volumes measured at room temperature and pressure)—

i.e. 2 volumes ammonia $\rightarrow$ 3 volumes hydrogen + 1 volume nitrogen.

Assuming hydrogen and nitrogen are diatomic (*i.e.* H_2 and N_2) then 2 volumes of ammonia $\rightarrow$ 3 volumes H_2 + 1 volume N_2. Using Avogadro's Law (number of molecules in a gas $\propto$ volume), then

$$\text{2 molecules of ammonia} \rightarrow 3H_2 + N_2$$

$\therefore$ 1 molecule ammonia contains 3H + N, *i.e.* formula = NH_3.

14 Metals(I)

14.1 Metals and non-metals

Most of the elements known to man are metals. The distinction between a metal and a non-metal is not, however, always clear. Elements that have properties in between metals and non-metals are called metalloids. Table 14.1 compares the properties of metals and non-metals.

Table 14.1 Comparison of properties of typical metals and non-metals

Metals	*Non-metals*
Solid at room temperature	Solid, liquid or gas at room temperature
Shiny	Dull
High density	Low density
Good conductor of heat and electricity	Poor conductor of heat and electricity
Can be beaten into thin sheets (malleable) and drawn into wire (ductile)	Brittle
Oxides are neutral or alkaline	Oxides are acidic or neutral
Forms positive ions (cations)	Forms negative ions (anions)
Sometimes forms hydrogen with dilute acids	Never forms hydrogen with dilute acids
Chlorides solid	Chlorides solid, liquid or gas

These distinctions between metal and non-metal are only generalisations. There are many exceptions. For example, mercury is a metal but is liquid at room temperature. Graphite is a form of carbon (non-metal) but it conducts electricity.

In this unit, the properties and uses of some common metals will be considered. The metals are potassium, sodium, calcium, magnesium, aluminium, zinc, iron, lead and copper. An outline of the methods of extraction of these metals from their ores and a consideration of the comparative reactivity of these metals will be found in unit 15.

14.2 Potassium K

Potassium is a reactive metal which is stored in oil to prevent it reacting with air or water. It belongs to the alkali metal family (see 18.2).

Potassium reacts with cold water (see 11.9). In all reactions a potassium atom readily loses a single electron and forms a potassium ion (K^+).

There are no uses for the metal but the compounds are widely used. Potassium nitrate KNO_3 (saltpetre) is a constituent of gunpowder and is preferred to the cheaper sodium nitrate because it is not deliquescent (see 11.7).

The compounds of potassium are very stable and are not easily decomposed by heating. Potassium sulphate, carbonate and hydroxide, do not decompose on heating and potassium nitrate only decomposes slightly.

$$2KNO_3(s) \rightarrow 2KNO_2(s) + O_2(g)$$
potassium nitrate → potassium nitrite + oxygen

All common potassium compounds are soluble in water.

14.3 Sodium Na

Sodium belongs to the alkali metal family and is also stored in oil to prevent reactions with air and water. A sodium atom (electronic structure 2,8,1) readily loses the electron in the third energy level and forms a sodium ion (Na^+).

Sodium reacts readily with cold water (see 11.9) and with air or oxygen.

With limited air or oxygen

$$4Na(s) + O_2(g) \rightarrow 2Na_2O(s)$$
sodium + oxygen → sodium monoxide
(white solid)

With excess air or oxygen

$$2Na(s) + O_2(g) \rightarrow Na_2O_2(s)$$
sodium + oxygen → sodium peroxide
(yellow solid)

Both oxides react rapidly with water producing alkaline solutions.

$$O^{2-}(s) + H_2O(l) \rightarrow 2OH^-(aq)$$
oxide ions + water → hydroxide ions

$$O_2^{2-}(s) + 2H_2O(l) \rightarrow 2OH^-(aq) + H_2O_2(l)$$
peroxide ion + water → hydroxide ion + hydrogen peroxide

Sodium peroxide is a very strong oxidising agent.

Sodium burns in chlorine gas producing sodium chloride (see 18.2).

Sodium hydroxide is a white deliquescent solid. It is strongly alkaline and is not decomposed by heating.

Sodium hydroxide readily absorbs carbon dioxide gas producing sodium carbonate and, further, sodium hydrogencarbonate.

$$2NaOH(aq) + CO_2(g) \rightarrow Na_2CO_3(aq) + H_2O(l)$$
sodium hydroxide + carbon dioxide → sodium carbonate + water

$$Na_2CO_3(aq) + H_2O(l) + CO_2(g) \rightarrow 2NaHCO_3(aq)$$
sodium carbonate + water + carbon dioxide → sodium hydrogencarbonate

Sodium hydroxide is produced by the electrolysis of sodium chloride solution (brine) in the **Kellner-Solvay process.** The anode (positive electrode) is made of graphite and the cathode is a moving stream of mercury passing over graphite blocks, see Fig. 14.1

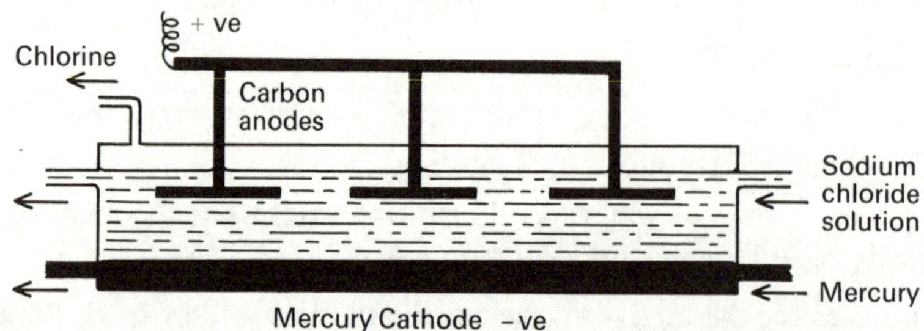

Fig. 14.1 Kellner-Solvay process (flowing mercury cathode)

At the anode, chlorine is produced from the discharge of chloride ions.

$$2Cl^-(aq) \rightarrow Cl_2(g) + 2e^-$$
chloride ions → chlorine + electrons lost to anode

At the cathode, sodium is discharged in preference to hydrogen ions, forming a sodium amalgam when it dissolves in the mercury. (This is an exception to the usual situation of hydrogen ions being discharged in preference to sodium ions.)

$$Na^+(aq) + e^- \rightarrow Na(s)$$
sodium ions + electron → sodium (which forms an amalgam with mercury)

When the sodium amalgam leaves the cell it enters a tank of water and sodium hydroxide and hydrogen are produced.

$$2Na_{(amalgam)} + 2H_2O(l) \rightarrow 2NaOH(aq) + H_2(g)$$
sodium amalgam + water → sodium hydroxide + hydrogen

The mercury is then re-cycled. It is, however, important to ensure that no mercury remains in the sodium hydroxide. This is because it is expensive and also very poisonous. The products of this process are sodium hydroxide, hydrogen and chlorine (see 42).

There are few large scale uses for sodium. However, it is a very powerful reducing agent and can be used to extract titanium metal from its chloride.

$$TiCl_4(l) + 4Na(l) \rightarrow Ti(s) + 4NaCl(s)$$
titanium(IV) chloride + sodium → titanium + sodium chloride

Sodium is also used as a coolant in some nuclear power stations.

Sodium compounds are very widely used. Sodium hydroxide (caustic soda) is used in the manufacture of soap, paper and aluminium (see 15.6). Sodium carbonate (soda ash) is used in glass making. Sodium hydrogencarbonate (sodium bicarbonate) is the agent that causes cakes to rise. The sodium hydrogencarbonate decomposes on heating to produce carbon dioxide which forms small bubbles within the cake.

$$2NaHCO_3(s) \rightarrow Na_2CO_3(s) + H_2O(g) + CO_2(g)$$
sodium hydrogencarbonate → sodium carbonate + water + carbon dioxide

The compounds of sodium closely resemble the compounds of potassium.

14.4 Calcium Ca

Calcium is commonly found in the earth usually as the carbonate or the sulphate. It is a reactive metal which corrodes rapidly, eventually forming calcium carbonate. Calcium belongs to the alkaline earth metal family (group II of the Periodic Table) and the two electrons in the outer energy level are easily lost to form calcium ions (Ca^{2+}). Other members of the same family include magnesium and barium.

Calcium reacts slowly with cold water (see 11.9) and rapidly with dilute acids.

E.g. $Ca(s) + 2HCl(aq) \rightarrow CaCl_2(aq) + H_2(g)$
calcium + hydrochloric acid → calcium chloride + hydrogen

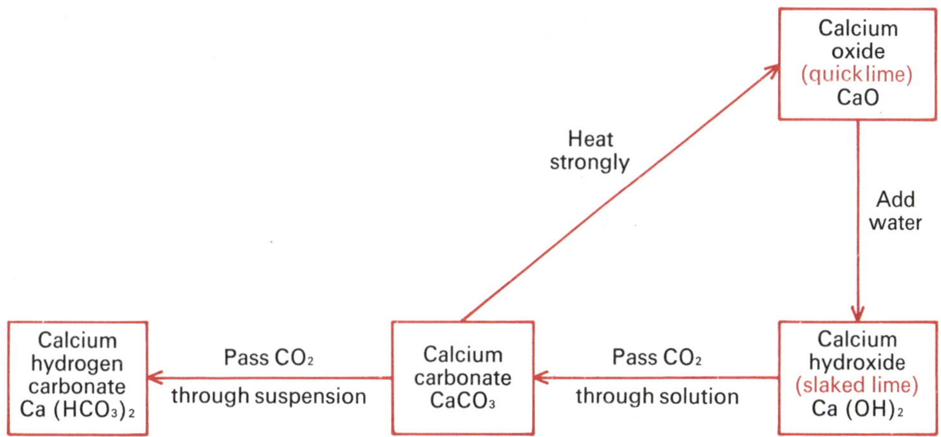

Fig. 14.2 Relationship between common calcium compounds

The relationship between common calcium compounds is shown in Fig. 14.2. The equations for these reactions are:

calcium carbonate → calcium oxide

$$CaCO_3(s) \rightarrow CaO(s) + CO_2(g)$$
calcium carbonate → calcium oxide + carbon dioxide

calcium oxide → calcium hydroxide

$$H_2O(l) + CaO(s) \rightarrow Ca(OH)_2(s)$$
water + calcium oxide → calcium hydroxide

calcium hydroxide → calcium carbonate

$$Ca(OH)_2(aq) + CO_2(g) \rightarrow CaCO_3(s) + H_2O(l)$$
calcium hydroxide + carbon dioxide → calcium carbonate + water
(limewater)

(This is the reaction taking place when carbon dioxide is passed into limewater and the solution goes milky, see 34.3.)

calcium carbonate → calcium hydrogencarbonate

$$CaCO_3(s) + H_2O(l) + CO_2(g) \rightleftharpoons Ca(HCO_3)_2(aq)$$
calcium carbonate + water + carbon dioxide ⇌ calcium hydrogencarbonate

(This is the reaction taking place when carbon dioxide is passed into limewater for a long time.)

Calcium metal has few uses. It is incorporated in some alloys, *e.g.* 3% of calcium is added to lead for bearings. However, calcium compounds are widely used. Calcium carbonate (limestone, chalk and marble) is used to make glass, cement, slaked lime, and is also used in the extraction of iron (see 15.7). Calcium sulphate is used to make plaster.

Calcium compounds are less stable to heat than compounds of potassium and sodium. Calcium carbonate and calcium nitrate are decomposed by heat.

$$CaCO_3(s) \rightarrow CaO(s) + CO_2(g)$$
calcium carbonate → calcium oxide + carbon dioxide

$$2Ca(NO_3)_2(s) \rightarrow 2CaO(s) + 4NO_2(g) + O_2(g)$$
calcium nitrate → calcium oxide + nitrogen dioxide + oxygen

Calcium compounds are not all soluble in water. Calcium carbonate is insoluble and calcium sulphate is sparingly soluble.

14.5 Magnesium Mg

Magnesium is also a member of the alkali metal family (group II of the Periodic Table). The metal corrodes more slowly than calcium. It burns in air or oxygen.

$$2Mg(s) + O_2(g) \rightarrow 2MgO(s)$$

magnesium + oxygen → magnesium oxide

But magnesium also reacts with nitrogen and so when magnesium reacts with air, a mixture of the nitride and the oxide are formed.

$$3Mg(s) + N_2(g) \rightarrow Mg_3N_2(s)$$

magnesium + nitrogen → magnesium nitride

Magnesium reacts very slowly with cold water but rapidly with steam (see 11.9). Fig. 14.3 shows apparatus suitable for demonstrating the reaction of magnesium with steam.

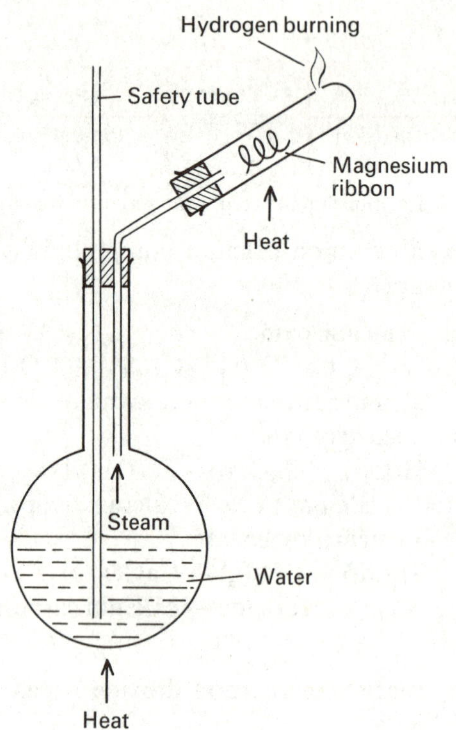

Fig. 14.3 Reaction of magnesium with steam

Magnesium reacts rapidly with dilute acids but is unaffected by alkalis.

$$E.g.\ Mg(s) + H_2SO_4(aq) \rightarrow MgSO_4(aq) + H_2(g)$$

magnesium + sulphuric acid → magnesium sulphate + hydrogen

Magnesium has a low density and it is used in lightweight alloys for aircraft. It is also used as a sacrificial metal to prevent iron rusting (see 15.9).

14.6 Aluminium Al

Aluminium is placed in group III of the Periodic Table. The electronic structure of an aluminium atom is 2,8,3 and it can form ions by losing three electrons (Al^{3+}). This requires a large amount of energy and aluminium frequently forms covalent compounds.

It has many typical metal properties (*e.g.* electrical conductivity) but it is not as reactive as expected. For example it reacts, only after a long initial delay, with warm, dilute hydrochloric acid.

$$2Al(s) + 6HCl(aq) \rightarrow 2AlCl_3(aq) + 3H_2(g)$$

aluminium + hydrochloric acid → aluminium chloride + hydrogen

The initial delay is caused by a layer of insoluble and unreactive aluminium oxide on the surface, which has to be removed before the reaction can start. The aluminium oxide can be removed by wiping with a cloth soaked in mercury(II) chloride or dipping the metal in mercury. When the oxide layer is removed, the aluminium is much more reactive. It is possible to make the oxide layer thicker by electrolysis, with the aluminium as the anode. Then the layer can be dyed using a suitable dye to give a very attractive finish for aluminium. This process is called **anodising**.

Aluminium reacts with dilute acids and alkalis.

Aluminium is a very strong reducing agent and can be used to reduce metal oxides, *e.g.* iron(III) oxide, chromium(III) oxide.

$$Fe_2O_3(s) + 2Al(s) \rightarrow Al_2O_3(s) + 2Fe(l)$$
iron(III) oxide + aluminium → aluminium oxide + iron

This is the basis of the Thermit process used for the welding of iron.

Aluminium is used for overhead power cables, car engines, aircraft construction and kitchen utensils.

Aluminium oxide has a high melting point which makes it suitable for refractory linings for kilns. It is an amphoteric oxide (see 9.6).

14.7 Zinc Zn

Zinc can be regarded as a heavy or transition metal (see 18.10) but it is not a typical transition metal. Atoms of zinc lose two electrons to form zinc ions (Zn^{2+}).

Zinc burns with a green flame when heated in air or steam.

$$2Zn(s) + O_2(g) \rightarrow 2ZnO(s)$$
zinc + oxygen → zinc oxide

$$Zn(s) + H_2O(g) \rightarrow ZnO(s) + H_2(g)$$
zinc + water → zinc oxide + hydrogen

Zinc reacts with dilute acids and hot alkalis producing hydrogen.

$$Zn(s) + H_2SO_4(aq) \rightarrow ZnSO_4(aq) + H_2(g)$$
zinc + sulphuric acid → zinc sulphate + hydrogen

$$Zn(s) + 2NaOH(aq) \rightarrow Na_2ZnO_2(aq) + H_2(g)$$
zinc + sodium hydroxide → sodium zincate + hydrogen

Zinc is widely used for galvanising iron to prevent corrosion. It is used to make alloys (*e.g.* brass) and for casings for dry batteries.

Zinc oxide is an amphoteric oxide. It is a white powder at room temperature but it turns yellow when heated due to the partial breaking up of the lattice. On cooling the original lattice reforms and a white powder remains.

14.8 Iron Fe

Iron is also a heavy or transition metal. It is the most widely used of all metals.

Iron (in a finely divided state) burns when heated in air or oxygen.

$$3Fe(s) + 2O_2(g) \rightarrow Fe_3O_4(s)$$
iron + oxygen → iron(II) di-iron(III) oxide

Iron rusts rapidly forming hydrated iron(III) oxide (see 15.9). Iron reacts reversibly with steam (see 11.9) and reacts with dilute acids (see 19.2). Iron does not react with alkalis.

Iron forms two ranges of compounds—iron(III) compounds, containing Fe^{3+} ions and iron(II) compounds, containing Fe^{2+} ions.

E.g. $FeCl_2$ iron(II) chloride $FeCl_3$ iron(III) chloride

Hydrated or aqueous solutions of iron(II) compounds are generally green and iron(III) compounds are yellow or brown.

Iron(II) chloride is produced when hydrogen chloride is passed over heated iron and iron(III) chloride is produced when chlorine is passed over heated iron (see 32.10).

$$Fe(s) + 2HCl(g) \rightarrow FeCl_2(s) + H_2(g)$$
iron + hydrogen chloride → iron(II) chloride + hydrogen

$$2Fe(s) + 3Cl_2(g) \rightarrow 2FeCl_3(s)$$
iron + chlorine → iron(III) chloride

Iron(II) compounds produce iron(III) compounds when oxidised, for example by chlorine.

$$2FeCl_2 + Cl_2(g) \rightarrow 2FeCl_3(aq)$$
iron(II) chloride + chlorine → iron(III) chloride

There are few uses for pure iron (wrought iron). It is used for decorative gates *etc.* Most iron is used in the form of the alloy, steel.

14.9 Lead Pb

Lead is the densest of the common metals. It is also very soft and has a low melting point for a metal. It is in group IV of the Periodic Table but shows no tendency to lose four electrons and form Pb^{4+} ions.

Lead is comparatively unreactive and reacts only slowly with acids, alkalis or air.

Lead is widely used for roofing, water pipes and radioactive sheilding.

Lead compounds are generally insoluble. (Only the nitrate and the ethanoate are soluble.) They are also inclined to decompose on heating, *e.g.* lead nitrate (see 28.6) and lead carbonate (see 34.5).

14.10 Copper Cu

Copper is a typical heavy or transition metal. It is not very reactive and does not react with water, alkalis or dilute acids (except dilute nitric acid).

$$3Cu(s) + 8HNO_3(aq) \rightarrow 3Cu(NO_3)_2(aq) + 4H_2O(l) + 2NO(g)$$
copper + nitric acid → copper nitrate + water + nitrogen monoxide

Copper forms a black coating of copper(II) oxide when heated in air.

$$2Cu(s) + O_2(g) \rightarrow 2CuO(s)$$
copper + oxygen → copper(II) oxide

Pure copper(II) oxide is not produced directly from copper and oxygen (see 9.2).
Copper reacts with concentrated nitric acid and sulphuric acid.

$$Cu(s) + 4HNO_3(l) \rightarrow Cu(NO_3)_2(aq) + 2H_2O(l) + 2NO_2(g)$$
copper + nitric acid → copper nitrate + water + nitrogen dioxide

$$Cu(s) + 2H_2SO_4(l) \rightarrow CuSO_4(aq) + 2H_2O(l) + SO_2(g)$$
copper + sulphuric acid → copper sulphate + water + sulphur dioxide

Copper has a large number of uses including making water pipes and electrical wiring. It is also used to make alloys, *e.g.* brass. When used for electrical purposes, it is important that it is as pure as possible. Otherwise its resistance is increased. It is purified by electrolysis.

The impure copper plate is made the anode (positive electrode) and a pure copper plate is used for the cathode (negative electrode). The electrolyte is copper(II) sulphate solution. During the electrolysis, copper dissolves from the anode.

$$Cu(s) \rightarrow Cu^{2+}(aq) + 2e^-$$
copper → copper ions + electrons

Pure copper is deposited on the cathode.

$$2e^- + Cu^{2+}(aq) \rightarrow Cu(s)$$
$$\text{electrons} + \text{copper ions} \rightarrow \text{copper}$$

At the bottom of the cell, 'anode slime' collects. This slime is rich in precious metals, *e.g.* silver and gold, and is formed from the impurities in the original impure copper.

14.11 Alloys

An **alloy** is composed of two or more metals mixed together. Sometimes a non-metal may be included, *e.g.* carbon in steel. By mixing the metals in this way, the product has more suitable properties when compared with a pure metal. Alloys are usually less malleable and ductile than pure metals. They also have lower melting points and electrical conductivity than pure metals. Some common alloys are included in Table 14.2.

An alloy is prepared by mixing the constituent metals in the correct proportions and melting the mixture.

Table 14.2 Examples of common alloys

Alloy	*Constituent elements*	*Uses*
Steel	Iron + between 0.15% and 1.5% carbon. The properties of steel depend on the percentage of carbon. Other metals may be present, *e.g.* chromium in stainless steel	Wide variety of uses including cars, ships, tools, reinforced concrete, tinplate (coated with tin)
Brass	Copper and zinc	Ornaments, buttons, screws
Duralumin	Aluminium, magnesium, copper and manganese	Lightweight uses *e.g.* aircraft, bicycles
Solder	Tin and lead	Joining metals (N.B. importance of low melting point)
Coinage bronze	Copper, zinc and tin	½p, 1p and 2p coins
Bronze	Copper and tin	Ornaments

15 Metals(II)

15.1 The reactivity series

In this unit, the comparative reactivity of the different metals is considered. Table 15.1 summarises the reactivity of the common metals.

Table 15.1 Reactions of some metals

Metal	*Metal heated in air*	*Metal with water or steam*	*Metal with acids*
Potassium	Burn to form oxide	React with cold water	Violent reaction with dilute acids
Sodium			
Calcium			
Magnesium		React with steam	React with dilute acids with decreasing ease
Aluminium			
Zinc			
Iron	React slowly	Reacts reversibly with steam	
Lead		Not attacked by water or steam	Attacked only by oxidising agents *e.g.* concentrated nitric acid
Copper			
Mercury	Reacts reversibly		
Silver	Do not react with air		
Platinum			No reaction

Following the detailed consideration of these reactions and other similar reactions, it is possible to arrange metals in a **reactivity series** (or **electrochemical series**). This is a list of metals in order of reactivity with the most reactive metal at the top of the list and steadily decreasing reactivity down the list (see left-hand column in Table 15.1).

Strictly speaking, the reactivity series (derived from the results of chemical reactions) is different from the electrochemical series. The electrochemical series is an arrangement of metals in order of standard electrode potentials. If two rods of different metals are dipped into dilute salt solution, a small voltage is produced which can be measured by means of a high resistance voltmeter. By comparing the voltages when different combinations of metals are used, it is possible to come up with essentially the same series. In this book, the term 'reactivity series' will be used.

15.2 Use of the reactivity series

In Chemistry, the reactivity series is widely used in order to
- (*i*) explain the stability of compounds when heated;
- (*ii*) explain the method used to extract a metal from its ore;
- (*iii*) explain the corrosion of certain metals, and
- (*iv*) predict possible reactions involving metals.

15.3 Stability of compounds

Compounds of metals high in the reactivity series are not easily split up by heating. Compounds of metals low in the reactivity series are split up by heating.

Examples of this can be seen by comparing the action of heat on carbonates (see 34.5) and nitrates (see 28.6).

This also explains why silver oxide is precipitated when sodium hydroxide is added to silver nitrate solution. The silver hydroxide expected splits up into silver oxide and water. (Silver is very low in the reactivity series.)

15.4 Extraction of metals from their ores

When answering any question on extraction of metals it is important to relate your method to the position of the metal in the reactivity series.

Some of the least reactive metals, *e.g.* gold, can be found in the form of the unreacted metal in the earth. Other metals can be found as compounds with other unwanted material in the form of an **ore**. Table 15.2 shows the ores of some common metals and the chief chemical constituent of the ore. (The examples in bold type are the ones most frequently asked for by examiners.)

Table 15.2 Common ores

Metal	*Ore*	*Chief chemical constituent*
Sodium	Rock salt	Sodium chloride
Calcium	Chalk, limestone, marble	Calcium carbonate
Magnesium	Magnesite also in sea water	Magnesium carbonate magnesium chloride
Aluminium	**Bauxite**	**Aluminium oxide**
Zinc	**Zinc blende**	**Zinc sulphide**
Iron	**Haematite**	**Iron(III) oxide**
Copper	Malachite	Basic copper(II) carbonate
Mercury	Cinnabar	Mercury(II) sulphide

Before the metal is extracted from the ore, the ore is frequently concentrated or purified.

Bauxite (aluminium oxide) is purified by adding the ore to sodium hydroxide solution. The aluminium oxide reacts and forms soluble sodium aluminate. Impurities, such as iron(III) oxide, can be removed by filtration. The aluminium oxide is then precipitated in a pure form (in fact aluminium hydroxide is precipitated and this is heated to give the pure oxide).

Zinc blende and galena can be concentrated by froth flotation. The ore is added to a detergent bath that is agitated. By careful control of the conditions in the bath, it is possible to cause the metallic sulphide to float and the impurities to sink.

The method used to extract the metal from the ore depends on the position of the metal in the reactivity series. If a metal is high in the reactivity series its ores are stable and the metal can only be obtained by electrolysis. Metals that can be obtained by electrolysis include potassium, sodium, calcium, magnesium and aluminium.

Metals in the middle of the reactivity series do not form very stable ores and they can be extracted by reduction, often with carbon. Examples of metals extracted by reduction are zinc, iron and lead.

Metals low in the reactivity series, if present in ores, can be extracted simply by heating because the ores are unstable. For example, mercury can be extracted by heating cinnabar.

Questions on extraction of metals often appear on examination papers and common examples are detailed below.

15.5 Extraction of sodium

Sodium is extracted by the electrolysis of molten sodium chloride in the Downs cell. Calcium chloride is added to the sodium chloride to lower the melting point to about 600°C. A Downs cell is shown in Fig. 15.1.

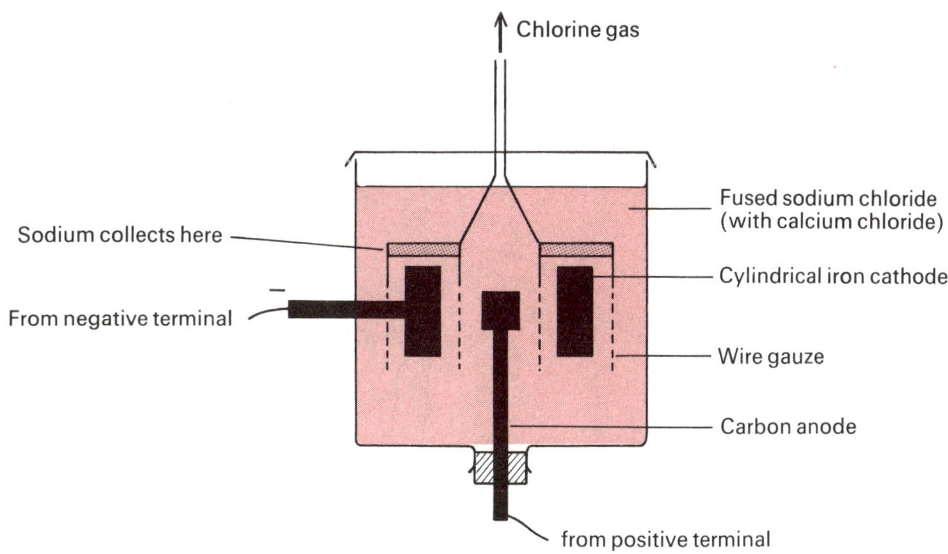

Fig. 15.1 Extraction of sodium

The cathode (negative electrode) is made of iron and the cylindrical anode (positive electrode) is made of graphite (carbon).

During the electrolysis the following reactions take place at the electrodes:

Cathode $Na^+(l) + e^- \rightarrow Na(l)$
Anode $2\,Cl^-(l) \rightarrow Cl_2(g) + 2e^-$

The sodium and chlorine produced are kept apart to prevent them reacting and reforming sodium chloride. The chlorine produced is a valuable by-product.

15.6 Extraction of aluminium

Aluminium is extracted from purified aluminium oxide (15.4) by electrolysis. However, aluminium oxide has a high melting point and is not readily soluble but it does dissolve in molten cryolite (Na_3AlF_6) and this produces a suitable electrolyte. An appropriate cell (Hall's cell) is shown in Fig. 15.2.

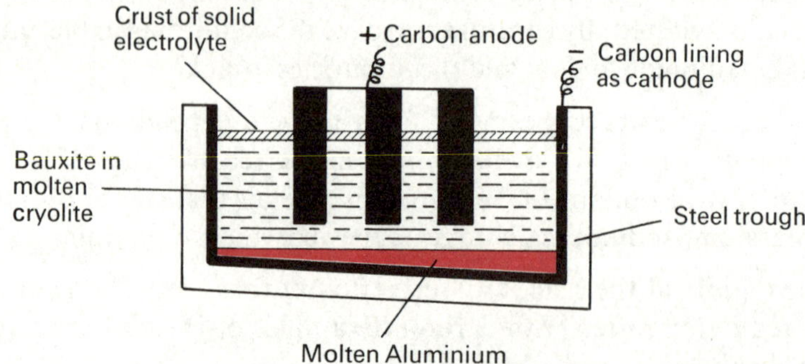

Fig. 15.2 Extraction of aluminium

The electrodes are made of carbon. The reactions taking place at the electrodes are as follows:

Cathode $Al^{3+} + 3e^- \rightarrow Al$

Anode $2\,O^{2-} \rightarrow O_2 + 4e^-$

At the working temperature of the cell, the oxygen reacts with the carbon of the anode to produce carbon dioxide. The anode has, therefore, to be replaced frequently. As this process requires a large amount of electricity, an inexpensive source, *e.g.* hydroelectric power, is an advantage.

15.7 Extraction of iron

Iron is extracted from iron ore, in large quantities, by reduction in a blast furnace (Fig. 15.3). The furnace is loaded with iron ore, coke and limestone and is heated by blowing hot air

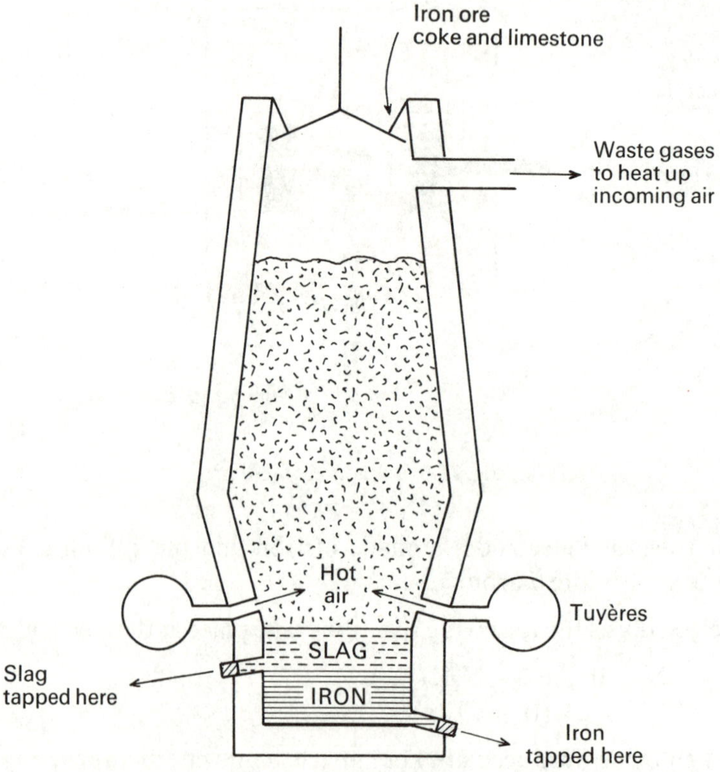

Fig. 15.3 Extraction of iron in a blast furnace

into the base from the Tuyères. Inside the furnace the following reactions take place raising the temperature to about 1500°C:

(*i*) The burning of the coke in the air

$$C(s) + O_2(g) \rightarrow CO_2(g)$$
carbon + oxygen → carbon dioxide

(*ii*) The reduction of the carbon dioxide to carbon monoxide

$$CO_2(g) + C(s) \rightarrow 2CO(g)$$
carbon dioxide + carbon → carbon monoxide

(*iii*) The reduction of the iron ore to iron by carbon monoxide

$$Fe_2O_3(s) + 3CO(g) \rightarrow 2Fe(l) + 3CO_2(g)$$
iron(III) oxide + carbon monoxide → iron + carbon dioxide

(*iv*) The decomposition of the limestone produces extra carbon dioxide

$$CaCO_3(s) \rightarrow CaO(s) + CO_2(g)$$
calcium carbonate → calcium oxide + carbon dioxide

(*v*) The removal of impurities by the formation of slag

$$CaO(s) + SiO_2(s) \rightarrow CaSiO_3(l)$$
calcium oxide + silicon dioxide → calcium silicate ('slag')

The molten iron sinks to the bottom of the furnace and the slag floats on the surface of the molten iron. The iron and slag can be tapped off separately at intervals. The iron produced is called pig iron and contains about 4% carbon. The slag, once discarded, is used as a phosphorus fertiliser and for road building.

Most of the pig iron is converted to steel. The usual method is to remove all the impurities by blowing air enriched with oxygen (or pure oxygen) through molten pig iron to produce pure iron (wrought iron). Calculated quantities of carbon and other substances can then be added to make the type of steel required.

15.8 Extraction of zinc

Following the concentration of the ore (see 15.4), the ore is heated strongly in air to convert the zinc sulphide to zinc oxide.

$$2ZnS(s) + 3O_2(g) \rightarrow 2ZnO(s) + 2SO_2(g)$$
zinc sulphide + oxygen → zinc oxide + sulphur dioxide

(The sulphur dioxide can be used to manufacture sulphuric acid, see 25.2.)

The zinc oxide and coke are mixed together in a cylindrical furnace and heated strongly. The carbon reduces zinc oxide to zinc.

$$ZnO(s) + C(s) \rightarrow Zn(g) + CO(g)$$
zinc oxide + carbon → zinc + carbon monoxide

At the temperature of the furnace, zinc distils off and is condensed. The carbon monoxide produced is used to heat the furnace.

15.9 Corrosion of metals

Iron is not the only metal to corrode. The corrosion of iron is so frequently quoted because of the economic importance of iron. Other metals corrode. Generally, the higher a metal is in the reactivity series, the more rapidly it corrodes. Potassium and sodium corrode rapidly and are, therefore, stored under oil to exclude air and water. The corrosion of some metals, *e.g.* aluminium and zinc, is prevented by protective oxide layers.

If a valuable steel object has to be protected from corrosion, *e.g.* a support of a pier which stands in sea water which would speed up the corrosion, it is possible to prevent corrosion by strapping a piece of magnesium (a more reactive metal) to the support in contact with the sea water. The magnesium corrodes in preference to the steel support. If a piece of copper (a less reactive metal) is used in place of magnesium, the corrosion is speeded up.

15.10 Predicting chemical reactions involving metals

A knowledge of the order of the metals in the reactivity series can be used to predict and explain chemical reactions involving metals.

The reactivity series can be written:

Potassium
Sodium
Calcium
Magnesium
Aluminium
(carbon)
Zinc
Iron
Lead
(hydrogen)
Copper
Mercury
Silver
Platinum

The inclusion of carbon and hydrogen in their correct places, although they are not metals, extends the usefulness of the reactivity series.

Metals above hydrogen will displace hydrogen from acids. The metals below will not displace hydrogen from dilute acids. Therefore, copper does not react with dilute hydrochloric acid.

Metals above carbon in the reactivity series are not produced by reduction of metal oxides with carbon. Metals below carbon can be produced by reduction of metal oxides with carbon.

If iron(III) oxide and aluminium powder are heated together, a reaction takes place because aluminium is more reactive than iron.

$$Fe_2O_3(s) + 2Al(s) \rightarrow 2Fe(s) + Al_2O_3(s)$$

iron(III) oxide + aluminium → iron + aluminium oxide

No reaction would take place if zinc oxide and copper were heated together because copper is less reactive than zinc.

If an iron nail is put into copper(II) sulphate solution, a reaction takes place. The blue solution turns almost colourless and brown copper is deposited.

$$Fe(s) + CuSO_4(aq) \rightarrow FeSO_4(aq) + Cu(s)$$

iron + copper(II) sulphate → iron(II) sulphate + copper

The reaction takes place because iron is more reactive than copper.

No reaction would take place if a piece of lead were put into a solution of magnesium sulphate.

Reactions of this type are called **displacement reactions** and it is important that the metal being added is more reactive than the metal already present in the compound if a reaction is to take place.

16 Oxidation and reduction

16.1 OXIDATION

Oxidation can be defined in various ways. These include when

(*i*) oxygen is added to a substance;
(*ii*) hydrogen is lost by a substance;
(*iii*) a substance loses electrons.

E.g. (*i*) If magnesium is burnt in oxygen, the magnesium is oxidised.

$$2Mg(s) + O_2(g) \rightarrow 2MgO(s)$$
magnesium + oxygen → magnesium oxide

When any substance burns in oxygen, it is oxidised.

E.g. (*ii*) If concentrated hydrochloric acid is oxidised, chlorine gas is produced.

$$MnO_2(s) + 4HCl(aq) \rightarrow MnCl_2(aq) + 2H_2O(l) + Cl_2(g)$$
manganese(IV) oxide + hydrochloric acid → manganese(II) chloride + water + chlorine

The concentrated hydrochloric acid loses hydrogen when being changed to chlorine and is therefore oxidised.

E.g. (*iii*) If chlorine is bubbled into a solution of iron(II) chloride (containing Fe^{2+} ions), the iron(II) chloride is oxidised to iron(III) chloride (containing Fe^{3+} ions).

$$2FeCl_2(aq) + Cl_2(g) \rightarrow 2FeCl_3(aq)$$
iron(II) chloride + chlorine → iron(III) chloride

or $$2Fe^{2+}(aq) + Cl_2(g) \rightarrow 2Fe^{3+}(aq) + 2Cl^-$$

Very frequently candidates confuse oxidation and reduction. Remember: *l*oss of *e*lectrons is *o*xidation **(leo).**

16.2 REDUCTION

Reduction is the reverse of oxidation. Reduction includes reactions where

(*i*) oxygen is lost by a substance;
(*ii*) hydrogen is gained by a substance;
(*iii*) a substance gains electrons.

E.g. (*i*) If hydrogen is passed over heated copper(II) oxide, the copper(II) oxide is reduced to copper. Copper(II) oxide loses oxygen.

$$CuO(s) + H_2(g) \rightarrow Cu(s) + H_2O(g)$$
copper(II) oxide + hydrogen → copper + water

E.g. (*ii*) If hydrogen and ethene are passed over a heated catalyst, the ethene is reduced to ethane. Ethene gains hydrogen.

$$C_2H_4(g) + H_2(g) \rightarrow C_2H_6(g)$$
ethene + hydrogen → ethane

E.g. (*iii*) During electrolysis of molten lead bromide (see 17.3), the lead ions are reduced to lead at the cathode (negative electrode)

$$Pb^{2+}(l) + 2e^- \rightarrow Pb(s)$$

Processes taking place at electrodes during electrolysis involve transfer of electrons and are oxidation or reduction.

16.3 REDOX REACTIONS

Oxidation and reduction processes occur together. If one substance gains electrons, another substance must lose electrons. A process where oxidation and reduction are taking place is called a redox (reduction–oxidation) reaction.

E.g. If a mixture of lead(II) oxide and carbon are heated together, the following reaction takes place:

$$PbO(s) + C(s) \rightarrow Pb(s) + CO(g)$$
lead(II) oxide + carbon → lead + carbon monoxide

In this reaction, lead(II) oxide is losing oxygen and forming lead. Lead(II) oxide is, therefore, being reduced. Carbon is gaining oxygen when it forms carbon monoxide. Carbon is being oxidised. Oxidation and reduction are both taking place.

No reaction would take place if the lead(II) oxide was heated alone. Carbon is the substance which is necessary for the reduction to take place because it removes the oxygen. Carbon is called the reducing agent. A reducing agent is a substance which reduces some other substance but it is itself oxidised.

Similarly, lead(II) oxide is the oxidising agent. It supplies oxygen which is used to oxidise the carbon. An oxidising agent is a substance which oxidises some other substance but is itself reduced.

16.4 COMMON OXIDISING AGENTS

These are substances that, by their presence, cause other substances to be oxidised.

Oxygen. During the reaction oxygen molecules (O_2) gain electrons and form oxide (O^{2-}) ions.

E.g.

$$2Mg(s) + O_2(g) \rightarrow 2MgO(s)$$
magnesium + oxygen → magnesium oxide

Chlorine. During the reaction, chlorine molecules (Cl_2) gain electrons and form chloride (Cl^- ions).

E.g.

$$Cl_2(g) + 2FeCl_2(aq) \rightarrow 2FeCl_3(aq)$$
chlorine + iron(II) chloride → iron(III) chloride

Potassium permanganate $KMnO_4$ (acidified with dilute sulphuric acid). During the reaction the solution turns from purple to become colourless as the permanganate (MnO_4^-) is reduced to manganese(II) ions.

E.g. with iron(II) sulphate solution

$$MnO_4^-(aq) + 8H^+(aq) + 5e^- \rightarrow Mn^{2+}(aq) + 4H_2O(l)$$
$$Fe^{2+}(aq) \rightarrow Fe^{3+}(aq) + e^-$$
$$MnO_4^-(aq) + 8H^+(aq) + 5Fe^{2+}(aq) \rightarrow Mn^{2+}(aq) + 4H_2O(l) + 5Fe^{3+}(aq)$$

Potassium dichromate $K_2Cr_2O_7$ (acidified with dilute sulphuric acid). During the reaction the solution turns from orange to green as the $Cr_2O_7^{2-}$ ion is reduced to Cr^{3+}.

E.g. with iron(II) sulphate solution

$$Cr_2O_7^{2-}(aq) + 14H^+(aq) + 6e^- \rightarrow 2Cr^{3+}(aq) + 7H_2O(l)$$
$$Fe^{2+}(aq) \rightarrow Fe^{3+}(aq) + e^-$$
$$Cr_2O_7^{2-}(aq) + 14H^+(aq) + 6Fe^{2+}(aq) \rightarrow 2Cr^{3+}(aq) + 7H_2O(l) + 6Fe^{3+}(aq)$$

Concentrated sulphuric acid H_2SO_4 (usually hot). When concentrated sulphuric acid acts as an oxidising agent, sulphur dioxide is always produced.

E.g. with copper

$$Cu(s) + 2H_2SO_4(l) \rightarrow CuSO_4(aq) + 2H_2O(l) + SO_2(g)$$
copper + sulphuric acid → copper(II) sulphate + water + sulphur dioxide

Concentrated nitric acid HNO_3 (usually hot). When concentrated nitric acid acts as an oxidising agent, nitrogen dioxide is produced.

E.g. with copper

$$Cu(s) + 4HNO_3(l) \rightarrow Cu(NO_3)_2(aq) + 2H_2O(l) + 2NO_2(g)$$
copper + nitric acid → copper(II) nitrate + water + nitrogen dioxide

16.5 Common reducing agents

These are substances that, by their presence, cause other substances to be reduced.

Hydrogen H_2

E.g.

$$C_2H_4(g) + H_2(g) \rightarrow C_2H_6(g)$$
ethene + hydrogen → ethane

Hydrogen sulphide H_2S. When hydrogen sulphide acts as a reducing agent, sulphur is always produced.

E.g. with chlorine

$$H_2S(g) + Cl_2(g) \rightarrow 2HCl(g) + S(s)$$
hydrogen sulphide + chlorine → hydrogen chloride + sulphur

Carbon C

E.g. with lead(II) oxide

$$PbO(s) + C(s) \rightarrow Pb(s) + CO(g)$$
lead(II) oxide + carbon → lead + carbon monoxide

Carbon monoxide CO. When carbon monoxide acts as a reducing agent, carbon dioxide is produced.

E.g. with iron(III) oxide (see 15.7)

$$Fe_2O_3(s) + 3CO(g) \rightarrow 2Fe(l) + 3CO_2(g)$$
iron(III) oxide + carbon monoxide → iron + carbon dioxide

Metals. For example, iron acts as a reducing agent with copper(II) sulphate solution (see 15.10).

$$Cu^{2+}(aq) + Fe(s) \rightarrow Fe^{2+}(aq) + Cu(s)$$
copper(II) ions + iron → iron(II) ions + copper

16.6 Substances which can act as both oxidising and reducing agents

Sulphur dioxide (in the presence of water) and hydrogen peroxide can act as both oxidising and reducing agents.

Sulphur dioxide, in the presence of water, forms sulphurous acid. When this acts as a reducing agent, it is oxidised to sulphuric acid.

E.g. with chlorine

$$SO_2(aq) + 2H_2O(l) + Cl_2(g) \rightarrow H_2SO_4(aq) + 2HCl(aq)$$

sulphur dioxide + water + chlorine → sulphuric acid + hydrochloric acid

In the presence of hydrogen sulphide (a stronger reducing agent), sulphur dioxide acts as an oxidising agent.

$$2H_2S(g) + SO_2(g) \rightarrow 3S(s) + 2H_2O(l)$$
hydrogen sulphide + sulphur dioxide → sulphur + water

Similarly, hydrogen peroxide can act as an oxidising agent or as a reducing agent.

E.g. With acidified potassium iodide solution (containing I^- ions), hydrogen peroxide acts an an oxidising agent and oxidises the iodide ions to iodine.

$$H_2O_2(aq) + 2H^+(aq) + 2e^- \rightarrow 2H_2O(l)$$
$$2I^-(aq) \rightarrow I_2(s) + 2e^-$$
$$\overline{H_2O_2(aq) + 2H^+(aq) + 2I^-(aq) \rightarrow 2H_2O(l) + I_2(s)}$$

With acidified potassium permanganate (a stronger oxidising agent than hydrogen peroxide), hydrogen peroxide acts as a reducing agent and reduces the permanganate ion to manganese(II) ion.

$$H_2O_2(aq) \rightarrow O_2(g) + 2H^+(aq) + 2e^-$$
$$MnO_4^-(aq) + 8H^+(aq) + 5e^- \rightarrow Mn^{2+}(aq) + 4H_2O(l)$$
$$5H_2O_2(aq) + 2MnO_4^-(aq) + 6H^+(aq) \rightarrow 5O_2(g) + 2Mn^{2+} + 8H_2O(l)$$

---

17 The effect of electricity on chemicals

This unit is written in two parts:

Part A 17.1–17.5 Qualitative electrolysis
Part B 17.6–17.7 Quantitative electrolysis

Check in Section I to find which parts are required for your syllabus

17.1 Conductors and insulators

A substance which allows electricity to pass through it is called a **conductor**. Of the solid elements at room temperature, only metals and graphite (a form of carbon) are good conductors. They conduct electricity because electrons pass freely through the solid.

Substances which do not allow electricity to pass through are called **insulators**. Some substances, e.g. germanium, conduct electricity slightly and are called **semiconductors** and they are important for making transistors.

17.2 Electrolytes

Certain substances do not conduct electricity when solid but conduct electricity when molten or dissolved in water. They are called **electrolytes**. However, the passage of electricity through the melt or solution is accompanied by a chemical decomposition. The splitting up of an electrolyte, when molten or in aqueous solution, is called **electrolysis**.

Electrolytes include:
acids, metal oxides, metal hydroxides and salts.

Electrolytes are composed of **ions** (*i.e.* ionic bonding, see 5.1) but in the solid state, the ions are rigidly held in regular positions and are unable to move to an electrode. Sodium chloride is composed of a regular lattice of sodium Na^+ and chloride Cl^- ions.

Melting the electrolyte breaks down the forces between the ions. The ions are, therefore, free to move in a molten electrolyte. In molten sodium chloride the sodium and chloride ions are able to move freely.

Dissolving an electrolyte in water (or other polar solvent) causes the breakdown of the lattice and the ions are free to move.

17.3 Electrolysis of molten lead bromide

Lead bromide is an electrolyte. The apparatus in Fig. 17.1 could be used for the electrolysis of molten lead bromide.

The bulb does not light up while the lead bromide is solid showing that no electric current is passing through the solid lead bromide. As soon as the lead bromide melts, the bulb lights up. After a while, bromine is seen escaping and, at the end of the experiment, lead can be found inside the crucible.

The reaction is, therefore

$$PbBr_2(l) \rightarrow Pb(l) + Br_2(g)$$
$$\text{lead bromide} \rightarrow \text{lead} + \text{bromine}$$

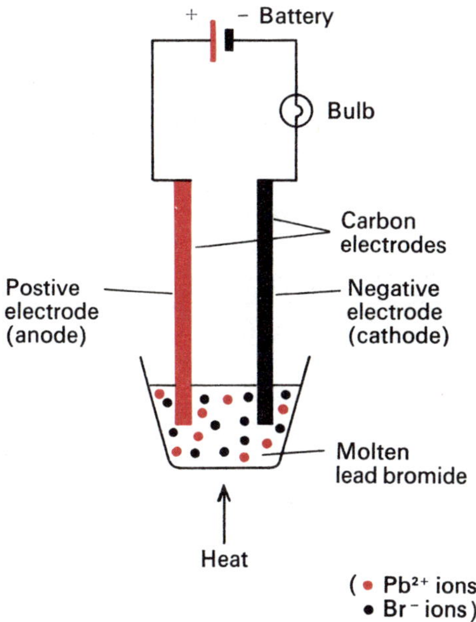

Fig. 17.1 Electrolysis of molten lead bromide

The carbon rods are called **electrodes**. The electrode attached to the positive terminal of the battery is the positive electrode (sometimes called the *anode*). The electrode attached to the negative terminal of the battery is the negative electrode (sometimes called the *cathode*). The positive electrode has a shortage of electrons and the negative electrode a surplus of electrons. Electrons are constantly flowing through the wire.

Lead bromide is composed of lead ions Pb^{2+} and bromide ions Br^-. When the lead bromide is molten, the ions move (or migrate) towards the electrode of opposite charge, *i.e.* Pb^{2+} ions to negative electrode and Br^- ions to positive electrode.

When the ions reach the oppositely charged electrode they are discharged. At the positive electrode (anode), bromide ions lose electrons to the electrode and form bromine molecules.

$$2Br^- \rightarrow Br_2 + 2e^- \qquad \text{(Oxidation)}$$

At the negative electrode (cathode), lead ions gain electrons from the electrode and form lead atoms.

$$Pb^{2+} + 2e^- \rightarrow Pb \qquad \text{(Reduction)}$$

The electrolysis of a molten electrolyte is comparatively easy to understand because only one positive and one negative ion are present. In a solution it is possible to have two positive ions and then these ions can compete to be discharged at the cathode.

17.4 Electrolysis of aqueous solutions

In pure water about 1 in every 600 000 000 water molecules ionise to form hydrogen and hydroxide ions.

$$H_2O(l) \rightleftharpoons H^+(aq) + OH^-(aq)$$

This very slight ionisation of water molecules explains the very slight electrical conductivity of pure water.

A solution of sodium chloride in water contains the following ions:

$H^+(aq)$	$OH^-(aq)$	from the water
$Na^+(aq)$	$Cl^-(aq)$	from the sodium chloride

Both positive ions migrate to the negative electrode and both negative ions to the positive electrode. At each electrode one or both of the ions may be discharged.

Table 17.1 below summarises the results of some electrolysis experiments.

Table 17.1 Examples of electrolysis of solutions

Solution	*Electrodes*	*Ion discharged at positive electrode*	*Ion discharged at negative electrode*	*Product at positive electrode*	*Product at negative electrode*
Dilute sulphuric acid	carbon	OH^-(aq)	H^+(aq)	oxygen	hydrogen
Dilute sodium hydroxide	carbon	OH^-(aq)	H^+(aq)	oxygen	hydrogen
Copper sulphate	carbon	OH^-(aq)	Cu^{2+}(aq)	oxygen	copper
Copper sulphate	copper	none	Cu^{2+}(aq)	none	copper
Copper(II) chloride	carbon	Cl^-(aq)	Cu^{2+}(aq)	chlorine	copper
Very dilute sodium chloride	carbon	OH^-(aq)	H^+(aq)	oxygen	hydrogen
Concentrated sodium chloride	carbon	Cl^-(aq)	H^+(aq)	chlorine	hydrogen
Concentrated sodium chloride	mercury cathode	Cl^-(aq)	Na^+(aq)	chlorine	sodium (amalgam)
Potassium iodide	carbon	I^-(aq)	H^+(aq)	iodine	hydrogen

The apparatus that could be used for the electrolysis of a solution and the collection of the gaseous products is shown in Fig. 17.2.

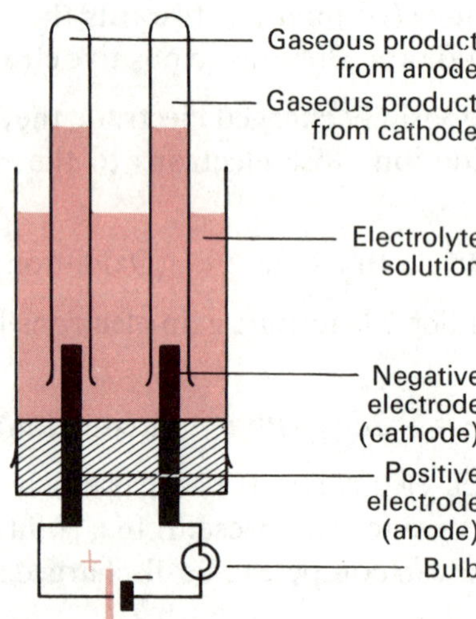

Fig. 17.2 Electrolysis of solutions

The following points about the electrolysis of solutions are worth remembering:

(*i*) Metals, if produced, are discharged at the negative electrode.

(*ii*) Hydrogen is produced at the negative electrode only.

(*iii*) Non-metals, apart from hydrogen, are produced at the positive electrode.

(*iv*) Reactive metals are not formed at the cathode during electrolysis of aqueous solutions. An exception is during the electrolysis of sodium chloride using a mercury cathode (see 14.3).

(*v*) The products obtained can depend upon the concentration of the electrolyte in the solution. For example, electrolysis of concentrated sodium chloride produces chlorine at the anode but electrolysis of dilute sodium chloride can produce oxygen at the anode.

(*vi*) Providing the concentrations of the negative ions in solution are approximately the same, the order of discharge is

$OH^-(aq)$
$I^-(aq)$
$Br^-(aq)$ — ease of discharge decreases ↓
$Cl^-(aq)$
$NO_3^-(aq)$
$SO_4^{2-}(aq)$

17.5 Uses of electrolysis

(i) Electroplating

Electrolysis can be used to form a very thin coating of a metal on the surface of another metal. This can be used for decorative purposes or to prevent corrosion.

If a piece of copper is to be nickel plated the surface must be clean and grease-free. The copper is made to be the cathode and a piece of nickel is the anode in an electrolysis experiment with nickel(II) sulphate solution the electrolyte. A nickel coating is deposited on the negative electrode (the copper) and the nickel anode goes into solution as nickel(II) ions.

(ii) Extraction of metals (see 15.4)
(iii) Purification of copper (see 14.10)
(iv) Manufacture of sodium hydroxide (see 14.3)

17.6 Quantitative electrolysis

If the bulb in Fig. 17.1 is replaced by an ammeter, it is possible to measure the current flowing in the circuit (in amps A) and to calculate the quantity of electricity passed in a certain time.

If a current of **1 A** flows for **1 second**, the quantity of electricity passed is **1 coulomb**. During an experiment it is found that, because a large number of coulombs are passed, it is better to work in units of Faradays.

1 Faraday (F) = 96 500 coulombs (approximately)

$$\text{Quantity of electricity passed} = \frac{\text{current (in A)} \times \text{time (in s)}}{96\,500}\ \text{F}$$

It is found, by experiment, that the quantity of electricity passed determines the mass of products formed.

Faraday's first law of electrolysis states that **the mass of a given element liberated during electrolysis is directly proportional to the quantity of electricity consumed during the electrolysis.**

17.7 Quantity of different elements deposited by the same quantity of electricity

Faraday's second law states that **the masses of different elements liberated by the same quantity of electricity, form simple whole number ratios when divided by their relative atomic masses.**

For example, during an electrolysis experiment, 1.08 g of silver are deposited and 0.32 g of copper. The relative atomic masses of silver and copper are 108 and 64 respectively. Therefore dividing the mass deposited by the appropriate relative atomic mass gives:

Silver	*Copper*
$\frac{1.08}{108} = 0.01$	$\frac{0.32}{64} = 0.005$

The answers obtained are in a simple ratio of 2:1. This is in accordance with Faraday's second law of electrolysis.

It is found that the discharge of one mole of atoms of an element from an ion with a single positive or negative charge requires 1 Faraday of electricity.

$\therefore$ 1 Faraday produces 108 g of silver from Ag^+ ions
or 35.5 g of chlorine from Cl^- ions

In a similar way, the discharge of one mole of atoms of an element from an ion with a double positive or negative charge requires 2 Faradays of electricity.

$\therefore$ 2 Faradays produces 64 g of copper from Cu^{2+} ions
or 16 g of oxygen from O^{2-} ions

Sample calculation
Calculate the mass of calcium atoms produced when a current of 5 A is passed for 32 min 10 s through molten calcium bromide ($Ca^{2+}.2Br^-$) (Ca = 40).

$$\text{Quantity of electricity passed} = 5 \times 1930 \text{ coulombs}$$
$$= \frac{5 \times 1930}{96\,500} \text{ Faradays}$$
$$= 0.1 \text{ Faradays}$$

Since calcium ions are Ca^{2+}, when 2 Faradays are passed, the number of moles of calcium atoms formed = 1.

$$\therefore 0.1 \text{ Faraday produces } \frac{0.1}{2} \text{ moles of calcium atoms}$$
$$= \frac{0.1}{2} \times 40 \text{ g of calcium}$$
$$= 2 \text{ g of calcium}$$

N.B. One Faraday of electricity contains 1 mole of electrons.

---

18 Chemical families and the Periodic Table

This unit is written in two parts:

Part A 18.1–18.4 Chemical families
Part B 18.5–18.10 Periodic Table

Check in Section I to find which parts are required for your syllabus

18.1 CHEMICAL FAMILIES

The 105 known elements can be divided into different groups in several ways. All elements can be classified as metals or non-metals and, from the appearance, it is easy to classify elements as solids, liquids or gases at room temperature. From a chemical point of view, it is useful to group elements together because they have similar chemical behaviour. These groups of elements are sometimes called families of elements. Elements in the same family are not identical but usually show marked similarities with other members of the same family.

18.2 THE ALKALI METAL FAMILY

This is a group of very reactive metals. The most common members of the family are lithium, sodium and potassium and some of their properties are shown in Table 18.1.

The metals have to be stored in oil to exclude air and water. They do not look much like metals, at first sight, but when freshly cut, they all have a typical shiny metallic surface.

They are also very good conductors of electricity. Note, however, that they have melting points and densities that are low compared to other metals.

Table 18.1

Element	*Symbol*	*Appearance*	*Melting point* (°C)	*Density* (g/cm³)
Lithium	Li	Soft grey metal	181	0.54
Sodium	Na	Soft light grey metal	98	0.97
Potassium	K	Very soft blue/grey metal	63	0.86

Reaction of alkali metals with water (see 11.9)
When a small piece of the alkali metal is put into a trough of water, the metal reacts immediately, floating on the surface of the water and evolving hydrogen.

With sodium and potassium, the heat evolved from the reaction is sufficient to melt the metal.

The hydrogen evolved by the reaction of potassium with cold water is usually ignited and burns with a pink flame.

Sodium reacts quicker than lithium and potassium reacts quicker than sodium.

In each case the solution remaining at the end of the reaction is an alkali.

$$2Li(s) + 2H_2O(l) \rightarrow 2LiOH(aq) + H_2(g)$$
lithium + water → lithium hydroxide + hydrogen
$$2Na(s) + 2H_2O(l) \rightarrow 2NaOH(aq) + H_2(g)$$
sodium + water → sodium hydroxide + hydrogen
$$2K(s) + 2H_2O(l) \rightarrow 2KOH(aq) + H_2(g)$$
potassium + water → potassium hydroxide + hydrogen

N.B. These three equations are basically the same and, if the alkali metal is represented by M, these equations can be represented by:

$$2M(s) + 2H_2O(l) \rightarrow 2MOH(aq) + H_2(g)$$

Reaction of alkali metals with oxygen
When heated in air or oxygen, the alkali metals burn to form white solid oxides. The colour of the flame is characteristic of the metal.

lithium—red
sodium—orange
potassium—lilac

E.g. $$4Li(s) + O_2(g) \rightarrow 2Li_2O(s)$$
lithium + oxygen → lithium oxide
or $$4M(s) + O_2(g) \rightarrow 2M_2O(s)$$

The alkali metal oxides all dissolve in water to form alkali solutions.

E.g. $$Li_2O(s) + H_2O(l) \rightarrow 2LiOH(aq)$$
lithium oxide + water → lithium hydroxide
or $$M_2O(s) + H_2O(l) \rightarrow 2MOH(aq)$$

Reaction of alkali metals with chlorine
When a piece of burning alkali metal is lowered into a gas jar of chlorine, the metal continues to burn forming a white smoke of the metal chloride.

E.g. $$2K(s) + Cl_2(g) \rightarrow 2KCl(s)$$
potassium + chlorine → potassium chloride
or $$2M(s) + Cl_2(g) \rightarrow 2MCl(s)$$

It is because of these similar reactions that these metals are put in the same family. In each reaction the order of reactivity is the same, *i.e.* lithium is least reactive and potassium is the most reactive.

There are three more members of this family—Rubidium Rb, Caesium Cs and Francium Fr. They are all more reactive than potassium.

18.3 THE HALOGEN FAMILY

This is a family of non-metals. In the alkali metal family, the members of the family all have similar appearances. In the halogen family, the different members have different appearances but they are put in the same family on the basis of their similar chemical reactions. Their appearances are compared in Table 18.2.

Table 18.2

Element	*Symbol*	*Appearance at room temperature*
Fluorine	F	Pale yellow gas
Chlorine	Cl	Yellow green gas
Bromine	Br	Red/brown volatile liquid
Iodine	I	Dark grey crystalline solid

There is another member of the family called Astatine. It is radioactive and a very rare element.

Fluorine is a very reactive gas and is too reactive to handle in normal laboratory conditions.

Solubility of halogens in water

None of the halogens is very soluble in water. Chlorine is the most soluble. Iodine does not dissolve much in cold water and only dissolves slightly in hot water.

Chlorine solution (sometimes called chlorine water) is very pale green. It turns Universal Indicator red (see 19.1) showing the solution is acidic (see 19.5). The colour of the indicator is quickly bleached.

Bromine solution (bromine water) is orange. It is very weakly acidic and also acts as a bleach.

Iodine solution is very weakly acidic and is also a slight bleach. The low solubility of halogens in water (a polar solvent) is expected because halogens are composed of molecules (see 5.7).

Solubility of halogens in hexane (a non-polar solvent)

The halogens dissolve readily in hexane to give solutions of characteristic colour.

chlorine	colourless
bromine	orange
iodine	purple

Reactions of halogens with iron

The halogens react with metals by direct combination to form salts. The name 'halogen' means salt producer. Chlorine forms chlorides, bromine forms bromides and iodine forms iodides.

If chlorine gas is passed over heated iron wire, an exothermic reaction takes place forming iron(III) chloride, which forms as a brown solid on cooling. Fig. 18.1 shows a suitable apparatus for preparing anhydrous iron(III) chloride crystals.

$$2Fe(s) + 3Cl_2(g) \rightarrow 2FeCl_3(s)$$
$$\text{iron} + \text{chlorine} \rightarrow \text{iron(III) chloride}$$

The anhydrous calcium chloride tube is to prevent water vapour entering the apparatus.

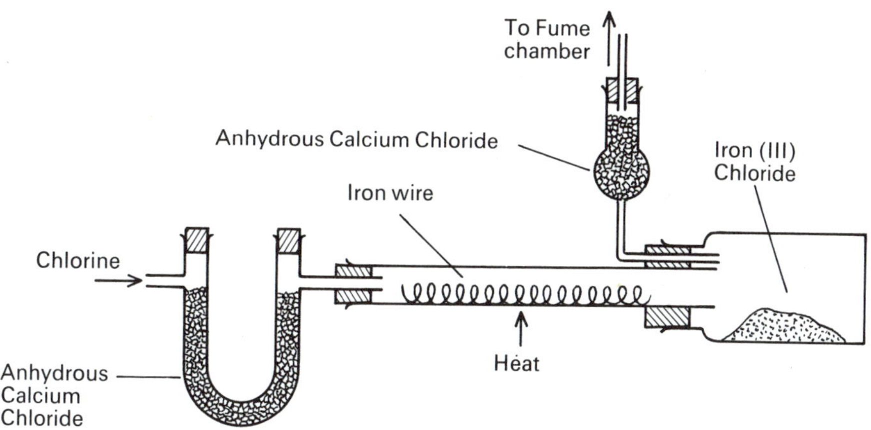

Fig. 18.1 Preparation of iron(III) chloride

Bromine vapour also reacts with hot iron wire to form iron(III) bromide. When iodine crystals are heated, they turn to a purple vapour. This vapour reacts with hot iron wire to produce iron(III) iodide.

Order of reactivity of the halogens

From their chemical reactions the relative reactivities of the halogens are:

Fluorine —most reactive
Chlorine
Bromine
Iodine —least reactive

Displacement reactions of the halogens

A more reactive halogen will displace a less reactive halogen from one of its compounds. For example when chlorine is bubbled into a solution of potassium bromide, the chlorine displaces the less reactive bromine. This means the colourless solution turns orange as the free bromine is formed.

$$2KBr(aq) + Cl_2(g) \rightarrow 2KCl(aq) + Br_2(aq)$$

potassium bromide + chlorine → potassium chloride + bromine

No reaction would take place if iodine solution was added to potassium bromide solution because iodine is less reactive than bromine.

18.4 THE NOBLE OR INERT GAS FAMILY

This is a family of gases. They are put in the same family because they are all very unreactive. Until about 20 years ago these gases were believed to be completely without chemical reactions. Since then a number of compounds, including xenon tetrafluoride XeF_4, have been produced. Table 18.3 gives some information about the noble gases.

Table 18.3

Element	*Symbol*	*Boiling point* (°C)	*Density* (g/dm³)
Helium	He	−269	0.17
Neon	Ne	−246	0.84
Argon	Ar	−185	1.66
Krypton	Kr	−153	3.46
Xenon	Xe	−109	5.45
Radon	Rn	−62	8.9

The noble gases are important because of their lack of reactions and they form the basis of theories of bonding.

Uses of the noble gases

Helium is used in balloons and airships. Although it is denser than hydrogen it has the advantage of not being inflammable.

Neon is used in advertising signs and **argon** is used to fill electric light bulbs. **Krypton** and **xenon** are used in lighthouse and projector bulbs.

Radon is radioactive and can be used to detect leaks in pipes.

18.5 THE PERIODIC TABLE

The Periodic Table is an arrangement of all the chemical elements in order of increasing atomic number with elements having similar properties (*i.e.* of the same chemical family) in the same vertical column. The Periodic Table is shown in Fig. 18.2 in a modern form.

18.6 BRIEF HISTORY OF THE DEVELOPMENT OF THE PERIODIC TABLE

In the early nineteenth century many new elements were being discovered and chemists were looking for similarities between these new elements and existing elements.

Döbereiner (1829) suggested that elements could be grouped in threes (triads). Each member of the triad has similar properties.

E.g. lithium, sodium, potassium
chlorine, bromine, iodine

Newlands (1863) arranged the elements in order of increasing relative atomic mass. He noticed that there was some similarity between each eighth element.

Li Be B C N O F
Na Mg Al Si P S Cl *etc.*

These were called **Newlands' Octaves**. Unfortunately the pattern broke down with the heavier elements and because he left no gaps for undiscovered elements. His work did not receive much support at the time.

Meyer (1869) looked at the relationship between relative atomic mass and the density of an element. He then plotted a graph of atomic volume (mass of 1 mole of atoms divided by density) against the relative atomic mass for each element. The curve he obtained showed periodic variations.

Mendeléff arranged the elements in order of increasing relative atomic mass but took into account the patterns of behaviour of the elements. He found it was necessary to leave gaps in the table and said that these were for elements not known at that time. His table enabled him to predict the properties of the undiscovered elements. His work was proved correct by the correct prediction of the properties of gallium and germanium. The Periodic Table we use today closely resembles the table drawn up by Mendeléff.

A modification of the Periodic Table was made following the work of **Rutherford** and **Moseley**. It was realised that the elements should be arranged in order of atomic number, *i.e.* the number of protons in the nucleus. In the modern Periodic Table the elements are arranged in order of increasing atomic number with elements with similar properties in the same vertical column.

18.7 STRUCTURE OF THE PERIODIC TABLE

The vertical columns in the table are called **groups**. A group will contain elements with similar properties. The groups are given Roman numbers as shown in Fig. 18.2.

The horizontal rows of elements are called **periods**.

The 'main block' elements are shaded in Fig. 18.2 and between the two parts of the main block are the heavy or transition metals.

	I	II	Transition elements										III	IV	V	VI	VII	O
1							1 H Hydrogen 1											4 He Helium 2
2	7 Li Lithium 3	9 Be Beryllium 4											11 B Boron 5	12 C Carbon 6	14 N Nitrogen 7	16 O Oxygen 8	19 F Fluorine 9	20 Ne Neon 10
3	23 Na Sodium 11	24 Mg Magnesium 12											27 Al Aluminium 13	28 Si Silicon 14	31 P Phosphorus 15	32 S Sulphur 16	35.5 Cl Chlorine 17	40 Ar Argon 18
4	39 K Potassium 19	40 Ca Calcium 20	45 Sc Scandium 21	48 Ti Titanium 22	51 V Vanadium 23	52 Cr Chromium 24	55 Mn Manganese 25	56 Fe Iron 26	59 Co Cobalt 27	59 Ni Nickel 28	64 Cu Copper 29	65 Zn Zinc 30	70 Ga Gallium 31	73 Ge Germanium 32	75 As Arsenic 33	79 Se Selenium 34	80 Br Bromine 35	84 Kr Krypton 36
	85.5 Rb Rubidium 37	88 Sr Strontium 38	89 Y Yttrium 39	91 Zr Zirconium 40	93 Nb Niobium 41	96 Mo Molybdenum 42	98 Tc Technetium 43	101 Ru Ruthenium 44	103 Rh Rhodium 45	106 Pd Palladium 46	108 Ag Silver 47	112 Cd Cadmium 48	115 In Indium 49	119 Sn Tin 50	122 Sb Antimony 51	128 Te Tellurium 52	127 I Iodine 53	131 Xe Xenon 54
	133 Cs Caesium 55	137 Ba Barium 56	139 La Lanthanum 57	178.5 Hf Hafnium 72	181 Ta Tantalum 73	184 W Tungsten 74	186 Re Rhenium 75	190 Os Osmium 76	192 Ir Iridium 77	195 Pt Platinum 78	197 Au Gold 79	201 Hg Mercury 80	204 Tl Thallium 81	207 Pb Lead 82	209 Bi Bismuth 83	210 Po Polonium 84	210 At Astatine 85	222 Rn Radon 86
	223 Fr Francium 87	226 Ra Radium 88	227 Ac Actinium 89															

KEY:

Atomic Mass
Symbol
Name
Atomic Number

139 La Lanthanum 57	140 Ce Cerium 58	141 Pr Praseodymium 59	144 Nd Neodymium 60	147 Pm Promethium 61	150 Sm Samarium 62	152 Eu Europium 63	157 Gd Gadolinium 64	159 Tb Terbium 65	162.5 Dy Dysprosium 66	165 Ho Holmium 67	167 Er Erbium 68	169 Tm Thulium 69	173 Yb Ytterbium 70	175 Lu Lutetium 71
227 Ac Actinium 89	232 Th Thorium 90	231 Pa Protactinium 91	238 U Uranium 92	237 Np Neptunium 93	242 Pu Plutonium 94	243 Am Americium 95	247 Cm Curium 96	247 Bk Berkelium 97	251 Cf Californium 98	254 Es Einsteinium 99	253 Fm Fermium 100	256 Md Mendeleevium 101	254 No Nobelium 102	257 Lw Lawrencium 103

Fig. 18.2 The Periodic Table of elements

18.8 Electron arrangement and reactivity in a group

In Unit 4 the arrangement of electrons within an atom was explained. The chemical properties of an element are controlled by the number of electrons in the outer energy level.

As elements in the same group have similar properties, we should expect some similarity in their electronic arrangement.

Table 18.4 shows the arrangement of electrons in the alkali metal family (group I of the Periodic Table).

Table 18.4

Element	*Atomic number*	*Arrangement of electrons*
Li	3	2,1
Na	11	2,8,1
K	19	2,8,8,1
Rb	37	2,8,18,8,1
Cs	55	2,8,18,18,8,1

Note that, in each case, the outer energy level contains just one electron. When an element reacts it attempts to obtain a full outer energy level.

Group I elements will lose one electron when they react and form a positive ion.

$$Na \rightarrow Na^{+} + e$$

We can explain the order of reactivity within the group. The electrons are held in position by the electrostatic attraction of the positive nucleus. This means that the closer the electron is to the nucleus, the harder it will be to remove it.

As we go down the group, the outer electron gets further away from the nucleus and so becomes easier to take away. This means as we go down the group, the reactivity should increase.

Table 18.5 shows the arrangement of electrons in the alkaline earth metal family (group II of the Periodic Table).

Table 18.5

Element		*Atomic number*	*Arrangement of electrons*
Beryllium	Be	4	2,2
Magnesium	Mg	12	2,8,2
Calcium	Ca	20	2,8,8,2
Strontium	Sr	38	2,8,18,8,2
Barium	Ba	56	2,8,18,18,8,2
Radium	Ra	88	2,8,18,32,18,8,2

As the atoms all have two electrons in their outer energy level, they will lose two electrons to form positive ions.

E.g.

$$Mg \rightarrow Mg^{2+} + 2e^{-}$$

More energy will be required to remove two electrons and so they will not be as reactive as the group I metals. As with group I, the reactivity will increase down the group.

E.g. Reaction of group II metals with water

Magnesium will react rapidly with steam.

$$Mg(s) + H_2O(g) \rightarrow MgO(s) + H_2(g)$$

magnesium + water (steam) → magnesium oxide + hydrogen

Calcium reacts with cold water.

$$Ca(s) + 2H_2O(l) \rightarrow Ca(OH)_2(aq) + H_2(g)$$
calcium + water → calcium hydroxide + hydrogen

Barium is stored in oil because it reacts rapidly with cold water.

$$Ba(s) + 2H_2O(l) \rightarrow Ba(OH)_2(aq) + H_2(g)$$
barium + water → barium hydroxide + hydrogen

Table 18.6 shows the arrangement of electrons in the halogen family (group VII of the Periodic Table).

Table 18.6

Element	*Atomic number*	*Arrangement of electrons*
F	9	2,7
Cl	17	2,8,7
Br	35	2,8,18,7
I	53	2,8,18,18,7

Note that each member of the group has seven electrons in the outer energy level. This is just one electron short of the full energy level.

When halogen elements react, they gain an electron to complete that outer energy level. This will form a negative ion.

E.g. $Cl + e^- \rightarrow Cl^-$

As an electron is being gained in the reaction, the most reactive member of the family will be the one where the extra electron is closest to the nucleus, *i.e.* fluorine. The reactivity decreases down the group.

18.9 Trends within a period

We have already seen that metallic elements form positive ions and that non-metals form negative ions. The metallic elements will then be those with only a few electrons in their outer energy level. The most metallic elements will be on the extreme left-hand side of the table, *i.e.* Group I.

The non-metallic elements will be on the right-hand side of the table. The heavy line in Fig. 18.2 divides metals from non-metals. In any period of the Periodic Table, there is a gradual change from metallic to non-metallic from left to right.

E.g. Third period
Na Mg Al Si P S O Cl Ne

Also from left to right in any period, the atoms gradually decrease in size. This surprises many people because, going from left to right, each element has one more electron than the previous element. However, this electron goes into the same energy level and the extra positive charge on the nucleus, caused by the extra proton, increases the attraction on the electrons and makes the atom slightly smaller.

18.10 Heavy or transition metals

The elements between groups II and III in the Periodic Table are called the heavy or transition metals. These metals are similar in many ways. They generally:

(*i*) have high melting points, high boiling points and high densities;
(*ii*) form more than one positive ion, *e.g.* iron forms Fe^{2+} and Fe^{3+} ions;
(*iii*) form many coloured compounds;
(*iv*) are often used as catalysts.

19 Acids, bases and salts

19.1 Acids

Acids form an important group of chemicals. They are generally thought of as being corrosive and having a sour taste.

Acids can be detected by using an **indicator** such as **litmus**. Indicators are dyes, or mixtures of dyes, which change colour when acids or alkalis are added. A wide range of indicators are available, each having its own characteristic colour in acid and alkali,

e.g. litmus — red in acid
blue in alkali

A very useful indicator is called **Universal Indicator**. This is a mixture of dyes and so gives a greater range of colour changes. The colours of Universal Indicator are shown in Fig. 19.1.

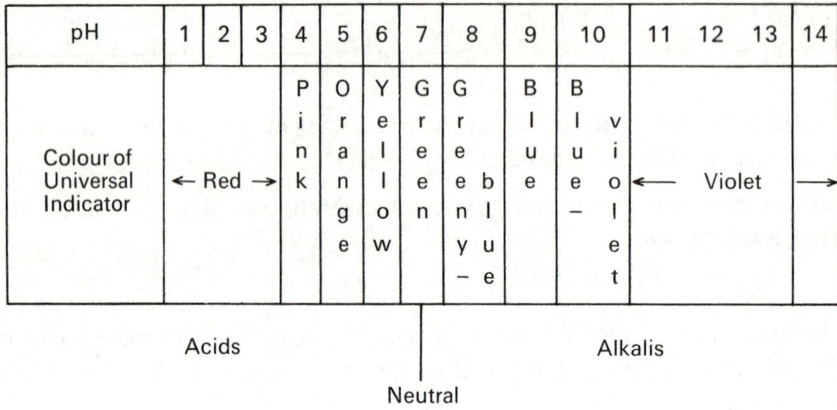

Fig. 19.1 Colours of Universal Indicator

Universal Indicator can show, not only whether a substance is acid, alkaline or neutral but can also show how strong the acid or alkali is. Table 19.1 gives a list of some common acids.

Table 19.1

Acid	*Formula*	
Hydrochloric acid	HCl	Mineral acids
Nitric acid	HNO_3	Mineral acids
Sulphuric acid	H_2SO_4	Mineral acids
Ethanoic acid (acetic acid)	CH_3COOH	Contained in vinegar
Citric acid	$C_6H_8O_7$	Contained in lemon juice

19.2 General reactions of acids

Although there are a large number of different acids, there are a number of general chemical reactions common to all acids.

(*i*) Acids turn indicators to their characteristic colour, *e.g.* litmus turns red.
(*ii*) Acids react with a fairly reactive metal to form a salt and evolve hydrogen.

E.g.
$$Mg(s) + H_2SO_4(aq) \rightarrow MgSO_4(aq) + H_2(g)$$
magnesium + sulphuric acid → magnesium sulphate + hydrogen

An exception is nitric acid. This acid tends to release oxides of nitrogen when reacted with metals, although very dilute nitric acid does liberate hydrogen with magnesium (see 28.5).

(*iii*) Acids react with metal oxides to form a salt and water. In most cases the acid needs warming.

E.g. $CuO(s) + 2HNO_3(aq) \rightarrow Cu(NO_3)_2(aq) + H_2O(l)$
copper(II) oxide + nitric acid → copper(II) nitrate + water

(*iv*) Acids react with metal carbonates to form a salt, carbon dioxide and water.

E.g. $CaCO_3(s) + 2HCl(aq) \rightarrow CaCl_2(aq) + CO_2(g) + H_2O(l)$
calcium carbonate + hydrochloric acid → calcium chloride + carbon dioxide + water

(*v*) Acids react with alkalis to form a salt and water.

E.g. $NaOH(aq) + HCl(aq) \rightarrow NaCl(aq) + H_2O(l)$
sodium hydroxide + hydrochloric acid → sodium chloride + water

19.3 Importance of Water

Hydrogen chloride gas will dissolve in both water and methylbenzene (toluene) but the solutions have different properties. Table 19.2 compares results of some tests for each solution.

Table 19.2

Test	*Solution of hydrogen chloride in water*	*Solution of dry hydrogen chloride in dry toluene*
(*i*) *Dry Universal Indicator paper*	Turns red showing strong acid	Turns green—neutral
(*ii*) *Add magnesium ribbon*	Hydrogen evolved	No reaction
(*iii*) *Add calcium carbonate*	Carbon dioxide evolved	No reaction
(*iv*) *Electrical conductivity*	Good conductor	Non-conductor
(*v*) *Temperature change on forming solution*	Rise in temperature	Little change

These results show that the solution in water behaves as an acid but the solution in methylbenzene shows no acidic properties.

The electrical conductivity of the solution in water indicates the presence of ions.

When hydrogen chloride gas dissolves in water it changes from molecules to ions.

$$HCl(g) \xrightarrow{\text{water}} H^+(aq) + Cl^-(aq)$$

This ionisation is accompanied by a rise in temperature. **It is the hydrogen ions that cause acidic properties and these can only be formed in the presence of water.**

You will come across different names and symbols for the hydrogen ion in solution which attempt to show how the hydrogen ion becomes associated with water molecules:

$H^+(aq)$ **hydrated proton** or **hydrated hydrogen ion**
$H_3O^+(aq)$ **oxonium** or **hydronium ion**

Remember that the hydrogen ion is simply a proton. (This is because a hydrogen atom consists of a single proton and a single electron; when H^+ is formed the electron is removed.)

19.4 Definitions of an Acid

An acid can be defined in various ways. These include:

(*i*) An acid is a substance that contains hydrogen which can be wholly or partially replaced by a metal. This hydrogen is called replaceable hydrogen.

(*ii*) An acid is a substance which forms hydrogen ions as the only positive ions when it is dissolved in water.

(*iii*) An acid is a proton donor. It provides protons or H^+ ions.

The general reactions of an acid (see 19.2) can now be seen as reactions of hydrogen ions in solution.

E.g.

$$Mg(s) + 2H^+(aq) \rightarrow Mg^{2+}(aq) + H_2(g)$$
$$O^{2-}(s) + 2H^+(aq) \rightarrow H_2O(l)$$
$$CO_3^{2-}(s) + 2H^+(aq) \rightarrow CO_2(g) + H_2O(l)$$
$$OH^-(aq) + H^+(aq) \rightarrow H_2O(l)$$

19.5 pH SCALE

As the reactions of an acid are those of the hydrogen ions in solution, clearly the more hydrogen ions present, the stronger the acid will be. The strength of an acid is measured on the **pH scale**. Fig. 19.1 relates the colour of Universal Indicator to the pH number. The pH number is a measure of the hydrogen ion concentration.

pH 1 strong acid
pH 2–pH 6 weaker acids
pH 7 neutral
pH 8–pH 13 weaker alkalis
pH 14 strong alkali

The smaller the pH the larger the concentration of hydrogen ions, and so a stronger acid. The pH can be measured with Universal Indicator or by using a pH meter.

19.6 STRONG AND WEAK ACIDS

Some acids completely ionise when they dissolve in water. These are called strong acids, *i.e.* the solution will contain a high concentration of hydrogen ions.

E.g. sulphuric acid

$$H_2SO_4(l) \xrightarrow{\text{water}} 2H^+(aq) + SO_4^{2-}(aq)$$

Other acids do not completely ionise on dissolving in water, *i.e.* some of the molecules remain un-ionised in the solution. These are called weak acids.

E.g. ethanoic acid (acetic acid)

$$CH_3COOH(l) \xrightleftharpoons{\text{water}} H^+(aq) + CH_3COO^-(aq)$$

In a molar solution of ethanoic acid only about four molecules in every thousand change into ions.

Table 19.3 lists some common strong and weak acids.

Table 19.3

Strong acids	*Weak acids*
Hydrochloric acid	Ethanoic acid
Sulphuric acid	Carbonic acid
Nitric acid	Sulphurous acid

19.7 BASES

A base is a substance that can accept hydrogen ions, *i.e.* a proton acceptor. A base will react with an acid to form a salt and water only.

E.g.

$$CuO(s) + H_2SO_4(aq) \rightarrow CuSO_4(aq) + H_2O(l)$$
copper(II) oxide + sulphuric acid → copper(II) sulphate + water
$$O^{2-}(s) + 2H^+(aq) \rightarrow H_2O(l)$$

Ionic equation:

If a base is soluble in water, the solution is called an **alkali**. Examples of alkalis include sodium hydroxide, NaOH, potassium hydroxide, KOH, and calcium hydroxide, $Ca(OH)_2$. An alkali contains a high concentration of hydroxide ions (OH^-) in solution.

When an acid reacts with an alkali, the reaction is between hydrogen ions and hydroxide ions.

$$H^+(aq) + OH^-(aq) \rightarrow H_2O(l)$$

This reaction is called a **neutralisation** reaction.

19.8 Strong and weak alkalis

If an alkali completely ionises on dissolving in water, a strong alkali is produced.

E.g. sodium hydroxide

$$NaOH(s) \xrightarrow{\text{water}} Na^+(aq) + OH^-(aq)$$

If an alkali does not completely ionise in water, a weak alkali is formed.

E.g. ammonium hydroxide (a solution of ammonia gas in water)

$$NH_4OH(aq) \rightleftharpoons NH_4^+(aq) + OH^-(aq)$$

19.9 Salts

A **salt** is a compound formed when the hydrogen of an acid is wholly or partly replaced by a metal.

The number of replaceable hydrogens in an acid is called the **basicity** of the acid. Table 19.4 gives the basicity of some common acids.

Table 19.4

Acid	*Formula*	*Basicity*
Hydrochloric acid	HCl	1
Nitric acid	HNO_3	1
Ethanoic acid	CH_3COOH	1
Sulphuric acid	H_2SO_4	2
Sulphurous acid	H_2SO_3	2
Carbonic acid	H_2CO_3	2
Phosphoric acid	H_3PO_4	3

In the case of an acid containing more than one replaceable hydrogen, salts can be formed when all or some of the hydrogen is replaced. Salts formed by replacing only part of the replaceable hydrogen are called **acid salts**.

Table 19.5 gives examples of salts formed by common acids.

Table 19.5

Acid	*Salt*	*Example*
Hydrochloric acid HCl	Chlorides	Sodium chloride NaCl
Nitric acid HNO_3	Nitrates	Sodium nitrate $NaNO_3$
Ethanoic acid CH_3COOH	Ethanoates (acetates)	Sodium ethanoate CH_3COONa
Sulphuric acid H_2SO_4	Sulphates	Sodium sulphate Na_2SO_4
	Hydrogensulphates (acid salt)	Sodium hydrogensulphate $NaHSO_4$
Carbonic acid	Carbonates	Sodium carbonate Na_2CO_3
	Hydrogencarbonates (acid salt)	Sodium hydrogencarbonate $NaHCO_3$

19.10 Preparation of salts

Before we can decide on the method of preparation of a salt, we need to know whether or not it is soluble in water.

The rules are as follows:

(*i*) All nitrates are soluble in water.
(*ii*) All sulphates are soluble in water except lead sulphate and barium sulphate. (Calcium sulphate is only slightly soluble in water.)
(*iii*) All chlorides are soluble in water except silver chloride, lead chloride and mercury(I) chloride.
(*iv*) All carbonates are insoluble in water except sodium carbonate, potassium carbonate and ammonium carbonate.
(*v*) All sulphides are insoluble except sodium sulphide, potassium sulphide and ammonium sulphide.
(*vi*) All salts of sodium, potassium and ammonium are soluble in water.

Both soluble and insoluble salts can be prepared by direct combination, *e.g.* iron(III) chloride (18.3), sodium chloride (18.2).

19.11 Preparation of soluble salts

For soluble salts, there are four general methods of preparation. These are the same reactions as the general reactions of an acid (see 19.2).

(i) Acid + metal
This method is only suitable for fairly reactive metals, *e.g.* magnesium, zinc and iron. The reaction is too vigorous for the more reactive metals. For metals below hydrogen in the reactivity series (see 15.10) this method is not suitable.

E.g. $Mg(s) + H_2SO_4(aq) \rightarrow MgSO_4(aq) + H_2(g)$
magnesium + sulphuric acid → magnesium sulphate + hydrogen

Magnesium powder is added to warm, dilute sulphuric acid in small amounts until excess magnesium powder remains in the solution. Excess magnesium powder is removed by filtering. Magnesium sulphate crystals are obtained by evaporating the solution until crystals form on the end of a glass rod, which was previously dipped into the hot solution. The solution is then left to cool.

(ii) Acid + metal oxide

This reaction requires warming to speed up the reaction. The method is the same as in (i).

E.g. $H_2SO_4(aq) + CuO(s) \rightarrow CuSO_4(aq) + H_2O(l)$
sulphuric acid + copper(II) oxide → copper(II) sulphate + water

(iii) Acid + metal carbonate

This reaction takes place at room temperature. The method is again the same as in (i).

E.g. $2HNO_3(aq) + CaCO_3(s) \rightarrow Ca(NO_3)_2(aq) + CO_2(g) + H_2O(l)$
nitric acid + calcium carbonate → calcium nitrate + carbon dioxide + water

(iv) Acid + alkali

This reaction requires a special technique as both reactants are solutions and so an indicator has to be used to show when reacting quantities of acid and alkali have been used.

Acid is added to a measured volume of alkali until the indicator changes colour. The process is then repeated using the same volumes of acid and alkali but without the indicator. The solution is then evaporated to obtain the salt.

E.g. $HCl(aq) + NaOH(aq) \rightarrow NaCl(aq) + H_2O(l)$
hydrochloric acid + sodium hydroxide → sodium chloride + water

19.12 PREPARATION OF ACID SALTS

To prepare an acid salt the exact amount of acid must be added to the alkali. If 25 cm^3 of sulphuric acid is required to produce the normal salt, sodium sulphate, then 50 cm^3 of sulphuric acid is required to produce the acid salt, sodium hydrogensulphate.

$$2NaOH(aq) + H_2SO_4(aq) \rightarrow Na_2SO_4(aq) + 2H_2O(l)$$
sodium hydroxide + sulphuric acid → sodium sulphate + water
$$NaOH(aq) + H_2SO_4(aq) \rightarrow NaHSO_4(aq) + H_2O(l)$$
sodium hydroxide + sulphuric acid → sodium hydrogensulphate + water

19.13 PREPARATION OF INSOLUBLE SALTS

There is only one method for preparing insoluble salts. This involves mixing together solutions of two soluble salts each containing half of the required salt. The required insoluble salt is then **precipitated**.

For example the insoluble salt lead carbonate can be prepared by mixing together solutions of a soluble lead salt (lead nitrate) and a soluble carbonate (sodium carbonate).

$$Pb(NO_3)_2(aq) + Na_2CO_3(aq) \rightarrow PbCO_3(s) + 2NaNO_3(aq)$$
lead nitrate + sodium carbonate → lead carbonate + sodium nitrate

Ionic equation:

$$Pb^{2+}(aq) + CO_3^{2-}(aq) \rightarrow PbCO_3(s)$$

This type of reaction is sometimes called **double decomposition** and is represented by the equation:

$$AX + BY \rightarrow AY + BX$$

In order to obtain a pure sample of the insoluble salt it is necessary to filter off the precipitate and then wash it with distilled water and finally dry the precipitate thoroughly.

20 Rates of reaction

20.1 INTRODUCTION

The **rate** of a chemical reaction is a measure of how fast the reaction takes place. It is important to remember that a rapid reaction is completed in a short time. Some reactions are very fast, *e.g.* the formation of silver chloride precipitate when silver nitrate and hydrochloric acid solutions are mixed. Other reactions are very slow, *e.g.* the rusting of iron. For practical reasons, reactions used in the laboratory for studying rates of reaction must not be too fast or too slow.

20.2 FINDING SUITABLE MEASUREABLE CHANGES

Having selected a suitable reaction it is necessary to find a change that can be observed during the reaction. An estimate of the rate of the reaction can be obtained from the time taken for the measurable change to take place. Suitable changes include:

(*a*) colour;
(*b*) formation of precipitate;
(*c*) change in mass (*e.g.* a gas evolved causing a loss of mass);
(*d*) volume of gas evolved;
(*e*) time taken for a given mass of reagent to disappear;
(*f*) pH;
(*g*) temperature.

20.3 STUDYING RATES OF REACTION

Some of the easiest reactions to study in the laboratory are those in which a gas is evolved. The reaction can be followed by measuring the volume of gas evolved over a period of time using the apparatus in Fig. 20.1.

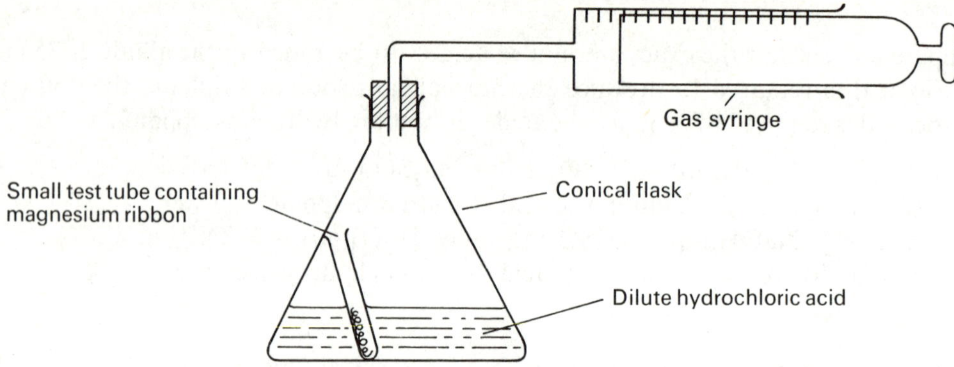

Fig. 20.1 Studying the reaction between magnesium and dilute hydrochloric acid

A suitable reaction is

$$Mg(s) + 2HCl(aq) \rightarrow MgCl_2(aq) + H_2(g)$$
$$\text{magnesium} + \text{hydrochloric acid} \rightarrow \text{magnesium chloride} + \text{hydrogen}$$

It is important to keep the reactants separate whilst setting up the apparatus so that the starting time of the reaction can be measured accurately.

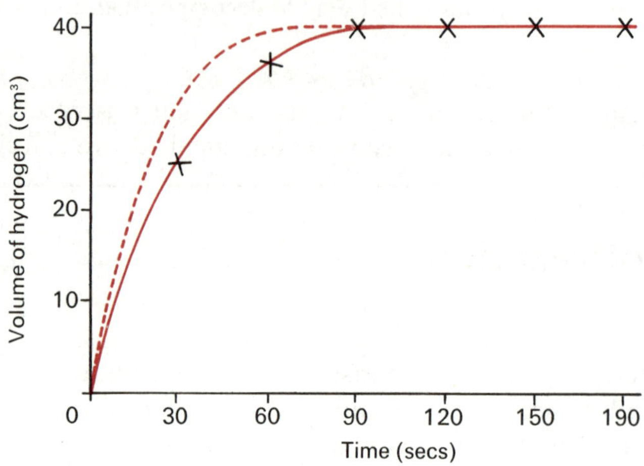

Fig. 20.2 A graph of volume of hydrogen collected at intervals

Fig. 20.2 shows a typical graph obtained for the reaction between dilute hydrochloric acid and magnesium. The dotted line shows the graph for a similar experiment using the same quantities of magnesium and hydrochloric acid but with conditions changed so that the reaction is slightly faster.

The rate of the reaction is greatest when the graph is steepest, *i.e.* at the start of the reaction. The reaction is complete when the graph becomes horizontal, *i.e.* there is no further increase in the volume of hydrogen.

It is often possible to follow the course of similar reactions by measuring the loss of mass during the reaction due to escape of gas. However, in this case, the loss of mass is very small.

20.4 Collision theory for rates of reaction

Before looking at the factors that can alter the rate of reaction, we must consider what happens when a reaction takes place.

First of all, the particles of the reacting substances must **collide** with each other and, secondly, a fixed amount of energy called the **activation energy (E_a)** must be reached if the reaction is to take place (Fig. 20.3).

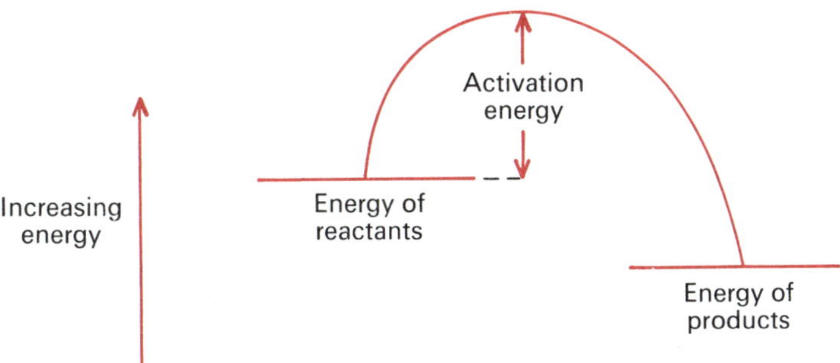

Fig. 20.3 Activation energy

If a collision between particles can produce sufficient energy (*i.e.* if they collide fast enough and in the right direction) a reaction will take place. Not all collisions will result in a reaction.

A reaction is speeded up if the number of suitable collisions is increased.

20.5 Effect of concentration on the rate of reaction

The more concentrated the reactants, the greater will be the rate of reaction. This is because increasing the concentration of the reactants increases the number of collisions between particles and, therefore, increases the rate of reaction.

This also explains why the greatest rate of reaction is usually as soon as the reactants are mixed *i.e.* they are both at their highest concentrations. As the reaction proceeds the concentrations of the reacting substances decrease and the rate of reaction decreases.

The effect of concentration can be shown by doing several experiments using equal masses of magnesium ribbon and hydrochloric acid of different concentrations.

Alternatively, a series of experiments can be carried out using a standard solution of sodium thiosulphate and hydrochloric acid solutions of different concentrations. The time is taken until the solution goes so cloudy that a cross disappears when viewed through the solution.

$$Na_2S_2O_3(aq) + 2HCl(aq) \rightarrow 2NaCl(aq) + S(s) + H_2O(l) + SO_2(g)$$

sodium thiosulphate + hydrochloric acid → sodium chloride + sulphur + water + sulphur dioxide

The cloudiness is due to the precipitation of sulphur.

20.6 Effect of pressure on the rate of reaction

When one or more of the reactants are gases an increase in pressure can lead to an increased rate of reaction. The increase in pressure forces the particles closer together. This causes more collisions and increases the rate of reaction.

20.7 Effect of temperature on the rate of reaction

An increase in temperature produces an increase in the rate of reaction. A rise of 10°C approximately doubles the rate of reaction.

When a mixture of substances is heated, the particles move faster. This has two effects. Since the particles are moving faster, they will travel a greater distance in a given time and so will be involved in more collisions. Also, because the particles are moving faster, a larger proportion of the collision will exceed the activation energy and so the rate of reaction increases.

The reaction between standard sodium thiosulphate and standard hydrochloric acid solutions at different temperatures can be used to examine the effect of temperature on the rate of reaction.

20.8 Effect of particle size on the rate of reaction

When one of the reactants is a solid, the reaction must take place on the surface of the solid. By breaking up the solid into smaller pieces, the surface area is increased, giving a greater area for collisions to take place and so causing an increase in the rate of reaction. This explains why mixtures of coal dust and air can cause explosions.

This effect can be examined by reacting equal masses of calcium carbonate with different particle sizes (*e.g.* chalk and marble chips) with equal volumes of the same hydrochloric acid solution.

$$CaCO_3(s) + 2HCl(aq) \rightarrow CaCl_2(aq) + H_2O(l) + CO_2(g)$$
calcium carbonate + hydrochloric acid → calcium chloride + water + carbon dioxide

20.9 Effect of light on the rate of reaction

The rates of some reactions are increased by exposure to light. Light has a similar effect, therefore, to increasing temperature.

Silver chloride, precipitated by mixing silver nitrate and hydrochloric acid solutions, turns from white to greyish-purple on exposure to sunlight due to the partial decomposition of silver chloride. The effects of light on hydrogen peroxide and concentrated nitric acid explain why they are stored in dark-glass bottles.

A mixture of hydrogen and chlorine does not react if kept in the dark, but in the presence of light an explosive reaction takes place.

20.10 Effect of catalysts on the rate of reaction

A **catalyst** is a substance which can alter the rate of a reaction but remains chemically unchanged at the end of the reaction. Catalysts usually speed up reactions. A catalyst which slows down a reaction is called a negative catalyst or **inhibitor**.

Catalysts speed up reactions by providing an alternative pathway for the reaction, *i.e.* one that has a much lower activation energy. More collisions will, therefore, have enough energy for this new pathway (Fig. 20.4).

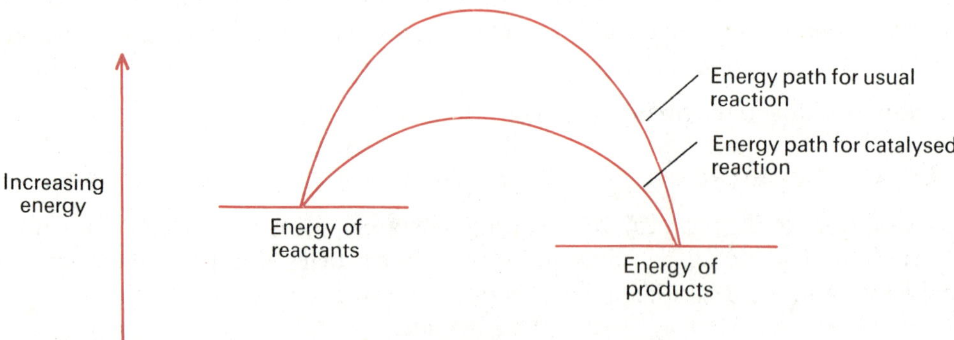

Fig. 20.4 Energy diagram for a catalysed reaction

In the laboratory the catalysed decomposition of hydrogen peroxide is usually studied.

$$2H_2O_2(aq) \rightarrow 2H_2O(l) + O_2(g)$$
hydrogen peroxide → water + oxygen

Catalysts for this reaction include manganese(IV) oxide and certain enzymes. **Enzymes** are biological catalysts.

Catalysts are important in many industrial processes. They do not increase the yield of the products but they do increase the rate of production.

Examples include:

(*i*) iron in the Haber process to produce ammonia (see 27.3);
(*ii*) vanadium(V) oxide in the Contact process to produce sulphuric acid (see 25.2);
(*iii*) platinum in the industrial conversion of ammonia to nitric acid (see 28.3).

Catalysts are usually heavy (transition) metals or compounds of heavy (transition) metals.

21 Reversible reactions and equilibrium

21.1 Introduction

Most chemical reactions can only go in one direction. For example, when magnesium is reacted with dilute hydrochloric acid, the products are hydrogen and magnesium chloride.

$$Mg(s) + 2HCl(aq) \rightarrow MgCl_2(aq) + H_2(g)$$
magnesium + hydrochloric acid → magnesium chloride + hydrogen

There is no way that the reverse reaction will take place. Hydrogen will not react with magnesium chloride, under any conditions, to produce magnesium and hydrochloric acid.

Some reactions, however, are reversible. A reversible reaction is a reaction that can go in either direction depending on the conditions of the reaction. There are a number of common examples of reversible reactions. The sign $\rightleftharpoons$ in an equation shows that the reaction is reversible.

21.2 Heating copper(II) sulphate crystals

When copper(II) sulphate crystals are heated, water vapour is driven off causing the blue crystals to turn to a white powder (anhydrous copper(II) sulphate). When a few drops of water are added to the cold white powder, the blue colour returns and heat is given out, showing the reversible nature of the reaction.

$$CuSO_4.5H_2O(s) \rightleftharpoons CuSO_4(s) + 5H_2O(l)$$
copper(II) sulphate crystals ⇌ anhydrous copper(II) sulphate + water

21.3 Heating ammonium chloride crystals

When ammonium chloride crystals are heated, the ammonium chloride dissociates into ammonia gas and hydrogen chloride gas. As the gases cool, they recombine to form solid ammonium chloride.

Similarly if the stopper from a bottle of concentrated ammonia solution is held near a stopper from a bottle of concentrated hydrochloric acid (evolving hydrogen chloride fumes), a dense white smoke of ammonium chloride is formed.

$$NH_4Cl(s) \rightleftharpoons NH_3(g) + HCl(g)$$

ammonium chloride ⇌ ammonia + hydrogen chloride

21.4 Formation of calcium hydrogencarbonate

Another readily reversible reaction involves the formation of calcium hydrogencarbonate.

If carbon dioxide is bubbled through a solution of calcium hydroxide (limewater), a cloudy white suspension of calcium carbonate is produced. This mixture goes clear again when more carbon dioxide is bubbled through, due to the formation of soluble calcium hydrogencarbonate.

Heating calcium hydrogencarbonate solution causes the reverse reaction and the cloudiness returns as insoluble calcium carbonate is reformed.

$$CaCO_3(s) + H_2O(l) + CO_2(g) \rightleftharpoons Ca(HCO_3)_2(aq)$$
calcium carbonate + water + carbon dioxide ⇌ calcium hydrogencarbonate

21.5 Reaction of iron with steam (see 11.9)

When steam is passed over heated iron, a slow reaction takes place producing hydrogen and an iron oxide. The apparatus suitable for this experiment is shown in Fig. 21.1.

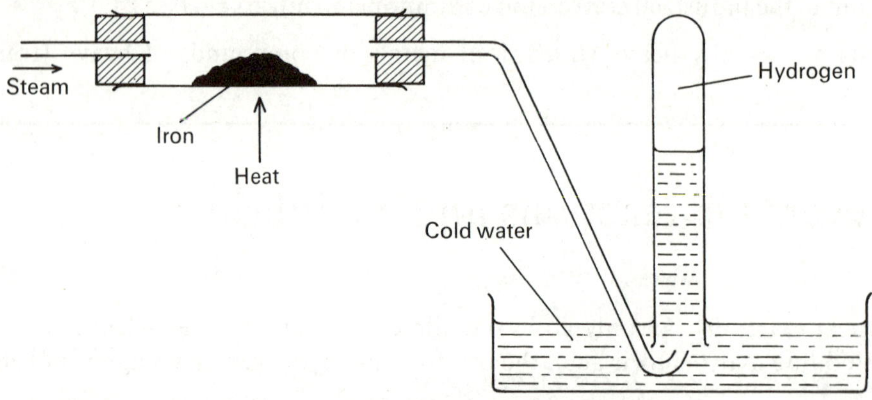

Fig. 21.1 Reaction of iron with steam

$$3Fe(s) + 4H_2O(g) \rightarrow Fe_3O_4(s) + 4H_2(g)$$

iron + water (steam) → iron(II) di-iron(III) oxide + hydrogen

In a second experiment dry hydrogen gas is passed over heated iron(II) di-iron(III) oxide (Fig. 21.2). This time the reverse reaction takes place and the iron oxide is reduced to iron.

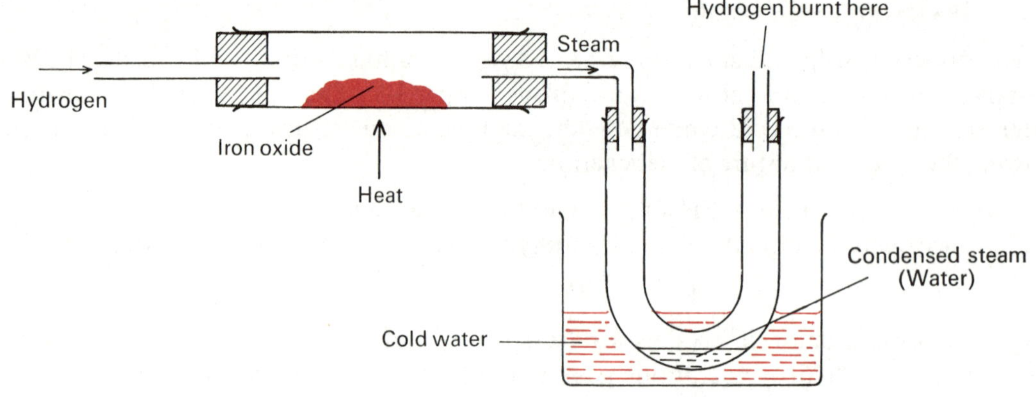

Fig. 21.2 Reaction of hydrogen and iron oxide

$$Fe_3O_4(s) + 4H_2(g) \rightarrow 3Fe(s) + 4H_2O(l)$$

An interesting consideration now is what would happen if the four chemicals were together present in a sealed container.

The iron would react with the steam to form hydrogen and iron oxide, but at the same time hydrogen would be reacting with the iron oxide to produce iron and steam.

$$Fe_3O_4(s) + 4H_2(g) \rightleftharpoons 3Fe(s) + 4H_2O(g)$$

After a time a state of balance would be set up where the rate of the forward reaction would be equal to the rate of the reverse reaction. This is called a state of **chemical equilibrium**.

Note that although there will appear to be no overall change in the concentrations of any of the chemicals, the equilibrium is a **dynamic** process *i.e.* both forward and reverse reactions are taking place but at the same rate.

An equilibrium can only be established in a sealed system, where no chemicals can enter or leave the system. A system in equilibrium is very delicately balanced. Any disturbance of the system, *e.g.* a change in temperature, may disturb the equilibrium by favouring the forward or reverse reaction.

Any reversible reaction can result in a chemical equilibrium, if the conditions are right, and so the sign $\rightleftharpoons$ can also mean a chemical equilibrium.

21.6 Factors affecting an equilibrium

In any equilibrium, the position of the equilibrium can be altered by changing the conditions.

Le Chatelier, a French chemist, stated a principle that governs the behaviour of equilibria. The principle states that **for a system in equilibrium, if any change is made to the conditions, the equilibrium will alter so as to oppose the change.**

Changes in conditions include temperature, pressure (if gases are involved) and concentration.

Effect of temperature
If the temperature of the system in equilibrium is lowered, the reaction will move in a direction to produce more heat, *i.e.* the exothermic reaction is favoured.

Effect of pressure
This applies to reactions involving gases. If the pressure is increased, the reaction will move to reduce the pressure by reducing the number of particles present.

Effect of concentration
If the concentration of one substance is increased, the reaction will move in a direction to use up the substance whose concentration was increased. If one substance is removed from the system, the reaction will move in a direction to produce more of the substance being removed.

The following two examples show how this principle can be applied to two important industrial processes.

21.7 Contact process for manufacture of sulphuric acid (see 25.2)

The important step in the manufacture of sulphuric acid is the reversible reaction:

$$2SO_2(g) + O_2(g) \rightleftharpoons 2SO_3(g) \qquad \Delta H = -385\text{ kJ}$$
$$\text{sulphur dioxide} + \text{oxygen} \rightleftharpoons \text{sulphur trioxide}$$

Effect of temperature
As the forward reaction is exothermic (ΔH is negative), lowering the temperature should cause the equilibrium to move to the right, producing more sulphur trioxide.

Effect of pressure
2 moles of sulphur dioxide molecules react with 1 mole of oxygen molecules to produce 2 moles of sulphur trioxide molecules. This means there will be a reduction in the number of molecules, and in the volume, as the forward reaction proceeds.

An increase in pressure will cause the equilibrium to move to the right.

Effect of concentration
Removal of the sulphur trioxide produced would cause the equilibrium to move to produce more sulphur trioxide.

An increase in the oxygen concentration will cause the equilibrium to move to reduce this, producing more sulphur trioxide.

Industrial conditions for the Contact process

(*i*) Atmospheric pressure. Although an increased pressure should improve the yield, about 98% of the gases can be converted without increasing pressure.
(*ii*) Temperature of 450°C.
(*iii*) Sulphur trioxide removed from the mixture.
(*iv*) Catalyst of vanadium(V) oxide (vanadium pentoxide) V_2O_5.

21.8 Haber process for the manufacture of ammonia (see 27.3)

Nitrogen reacts with hydrogen to form ammonia. This is a reversible reaction and under normal conditions the equilibrium lies very much to the left, meaning that very little ammonia gas is formed.

$$N_2(g) + 3H_2(g) \rightleftharpoons 2NH_3(g) \quad \Delta H = -91\ kJ$$
$$\text{nitrogen} + \text{hydrogen} \rightleftharpoons \text{ammonia}$$

Effect of temperature
The forward reaction is exothermic. If the system is cooled, the equilibrium will move to oppose this change, *i.e.* move to the right producing more ammonia.

Effect of pressure
From the equation it can be seen that 1 mole of nitrogen molecules reacts with 3 moles of hydrogen molecules to give 2 moles of ammonia molecules. This means that the formation of ammonia is accompanied by a reduction in the number of molecules. Increasing the pressure favours the conversion to ammonia.

Effect of concentration
If the concentration of one of the chemicals is altered, the equilibrium will move to oppose the change. This means, if the ammonia is removed from the system as it is formed, the equilibrium will move to the right to produce more ammonia.

Industrial conditions for the Haber process

(*i*) Very high pressure (about 200 atmospheres although some modern plants use pressures up to 1000 atmospheres).
(*ii*) Temperature of about 500°C. Note that a lower temperature would cause a greater proportion of ammonia to be formed but the rate of reaction is too slow.
(*iii*) Ammonia is removed as it is produced.
(*iv*) A catalyst of iron. This enables the equilibrium to be established more quickly but does not alter the position of the equilibrium.

---

22 Sulphur

22.1 Occurrence of sulphur

Sulphur is found uncombined in volcanic regions of Italy, including Sicily. It is also found in underground deposits in Texas and Louisiana (USA).

Sulphur is found in a wide range of compounds including:

(*i*) metallic sulphides, *e.g.* ZnS zinc blende, FeS_2 iron pyrites;
(*ii*) metallic sulphates, *e.g.* $CaSO_4$ anhydrite;
(*iii*) hydrogen sulphide present in certain natural gases.

22.2 Extraction of sulphur

The underground sulphur deposits in Texas and Louisiana cannot be mined in the usual way because of layers of quicksand between the surface and the deposits and because of the poisonous gases associated with the sulphur deposits.

An ingenious method devised by **Frasch** and called the **Frasch process** is now widely used. A hole is drilled from the surface to the deposits using an oil drill. The pump, consisting of three tubes one inside another, is inserted into the hole (Fig. 22.1).

Water, superheated to 170°C under pressure, is pumped down the outer pipe and this melts the sulphur. Hot compressed air is pumped down the middle pipe and this forces the mixture of molten sulphur and water to the surface.

The sulphur obtained by this method is 99.5% pure. It is sold in the form of cylinders and is called roll sulphur.

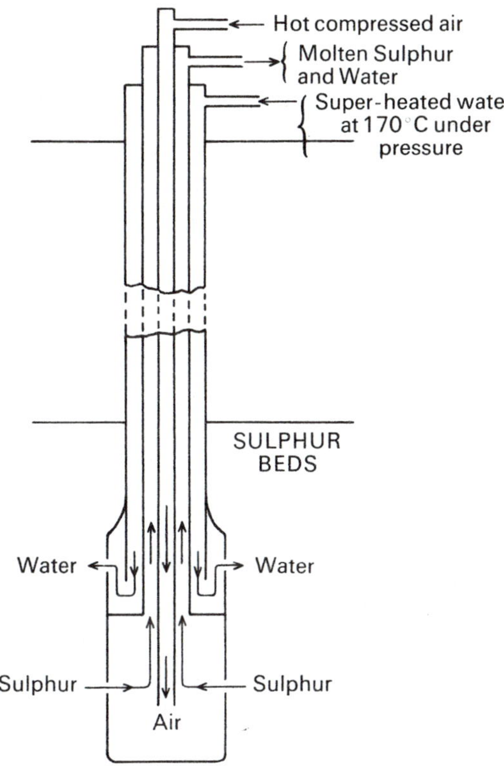

Fig. 22.1 Extraction of sulphur by the Frasch process

22.3 ALLOTROPES OF SULPHUR

Sulphur can exist in different allotropic forms. The stable allotrope of sulphur below 96°C is called α **(octahedral or rhombic) sulphur**. Above 96°C, the stable allotrope of sulphur is *β* **(prismatic or monoclinic) sulphur**. This temperature (96°C) is called the **transition temperature**. Both allotropes are stable at this temperature.

Both α and *β* sulphur are made up from regular but different arrangement of rings. Each ring contains eight sulphur atoms.

22.4 PREPARATION OF α SULPHUR

A solution of sulphur in xylene or carbon disulphide (CS_2) is prepared. The solution is allowed to evaporate at room temperature and pale yellow octahedral crystals are formed. The α sulphur is formed because the temperature is below 96°C throughout.

22.5 PREPARATION OF *β* SULPHUR

Sulphur is heated until molten and then allowed to cool. Brown needle-like crystals are found beneath the crust which forms on the surface. The crystals form above 96°C and so *β* sulphur is formed. The *β* sulphur changes to α sulphur when kept at room temperature (or any temperature below 96°C).

22.6 ACTION OF HEAT ON SULPHUR

When sulphur is heated, it melts at 119°C to form an amber-coloured free-running liquid. The free-running liquid consists of S_8 rings.

As the temperature is raised, the liquid first turns red and then black. It also becomes viscous (like treacle). The S_8 rings open out to form long chains of sulphur atoms.

On further heating, the liquid becomes less viscous and finally boils at 444°C. The long chains break down to form shorter chains.

When molten sulphur is poured into cold water, a brown rubbery material is formed called **plastic sulphur**. Plastic sulphur reverts to α sulphur when kept at room temperature.

22.7 Properties of sulphur

(*i*) Sulphur burns with a blue flame in air or oxygen to form sulphur dioxide.

$$S(s) + O_2(g) \rightarrow SO_2(g)$$
sulphur + oxygen → sulphur dioxide

(N.B. Sulphur trioxide is not formed when sulphur burns in excess air.)

(*ii*) Metallic sulphides are formed when a mixture of a metal powder and sulphur are heated.

E.g.
$$Fe(s) + S(s) \rightarrow FeS(s)$$
iron + sulphur → iron(II) sulphide

(*iii*) Sulphur is not attacked by dilute acids. It reacts with hot concentrated nitric or sulphuric acids (both oxidising agents).

$$S(s) + 2H_2SO_4(l) \rightarrow 3SO_2(g) + 2H_2O(g)$$
sulphur + sulphuric acid → sulphur dioxide + water

The sulphur is oxidised to sulphur dioxide and the sulphuric acid is reduced to sulphur dioxide and water.

$$S(s) + 6HNO_3(l) \rightarrow H_2SO_4(l) + 2H_2O(g) + 6NO_2(g)$$
sulphur + nitric acid → sulphuric acid + water + nitrogen dioxide

22.8 Uses of sulphur

Most of the sulphur and sulphur containing minerals are used in the manufacture of sulphuric acid (see 25.2).

Sulphur is also used:
(*i*) for hardening or vulcanising rubber;
(*ii*) producing calcium hydrogensulphite for bleaching paper pulp;
(*iii*) making medicines containing sulphur.

---

23 Sulphides and hydrogen sulphide

23.1 Sulphides

Sulphides are compounds of an element with sulphur. They are usually prepared by heating a mixture of the element and sulphur together. The formation of a compound from its constituent elements is called **synthesis**.

E.g. Iron(II) sulphide is formed by heating a mixture of iron and sulphur powders (ratio approximately 2:1 by mass).

$$Fe(s) + S(s) \rightarrow FeS(s)$$
iron + sulphur → iron(II) sulphide

The iron(II) sulphide produced often contains iron as an impurity.

23.2 Laboratory preparation of hydrogen sulphide

Hydrogen sulphide is prepared by the action of dilute hydrochloric acid on iron(II) sulphide.

$$FeS(s) + 2HCl(aq) \rightarrow FeCl_2(aq) + H_2S(g)$$
iron(II) sulphide + hydrochloric acid → iron(II) chloride + hydrogen sulphide

The apparatus used is the same as for the preparation of hydrogen (Fig. 10.1) except hydrogen sulphide is collected over warm water because it is quite soluble in cold water.

If hydrogen sulphide is required dry, it is dried by passing over phosphorus pentoxide (P_2O_5) and collected by downward delivery. It cannot be dried by passing through concentrated sulphuric acid as they react.

The chief impurity in the hydrogen sulphide produced by this method is hydrogen formed by the reaction between iron (present as impurity in the iron(II) sulphide) and hydrochloric acid.

$$Fe(s) + 2HCl(aq) \rightarrow FeCl_2(aq) + H_2(g)$$
iron + hydrochloric acid → iron(II) chloride + hydrogen

23.3 PROPERTIES OF HYDROGEN SULPHIDE

Hydrogen sulphide is a colourless gas which smells strongly of bad eggs.

(i) Combustion in air or oxygen (oxidation)

Hydrogen sulphide burns in air or oxygen with a blue flame. In a limited supply of air, the products are sulphur and water.

$$2H_2S(g) + O_2(g) \rightarrow 2H_2O(g) + S(s)$$
hydrogen sulphide + oxygen → water + sulphur

In excess air, the hydrogen sulphide burns to form sulphur dioxide and water.

$$2H_2S(g) + 3O_2(g) \rightarrow 2H_2O(g) + 2SO_2(g)$$
hydrogen sulphide + oxygen → water + sulphur dioxide

(ii) Precipitation reactions

Hydrogen sulphide gas, when passed through metallic salt solutions, can precipitate metallic sulphides.

E.g. When hydrogen sulphide is passed through lead nitrate solution, a black precipitate of lead(II) sulphide is formed.

$$Pb(NO_3)_2(aq) + H_2S(g) \rightarrow PbS(s) + 2HNO_3(aq)$$
lead(II) nitrate + hydrogen sulphide → lead(II) sulphide + nitric acid

If a filter paper soaked in lead(II) nitrate solution (or lead ethanoate) is put into hydrogen sulphide, the filter paper turns black. This is used as a test for hydrogen sulphide.

When hydrogen sulphide is passed through copper(II) sulphate solution, a black precipitate of copper(II) sulphide is formed.

$$CuSO_4(aq) + H_2S(g) \rightarrow CuS(s) + H_2SO_4(aq)$$
copper(II) sulphate + hydrogen sulphide → copper(II) sulphide + sulphuric acid

(iii) Hydrogen sulphide as a reducing agent

When hydrogen sulphide acts as a reducing agent, it is oxidised to sulphur.

$$H_2S \rightarrow S + 2H^+ + 2e^-$$

E.g. When hydrogen sulphide is passed into yellow iron(III) chloride solution, the iron(III) chloride is reduced to green iron(II) chloride.

$$H_2S(g) \rightarrow S(s) + 2H^+(aq) + 2e^-$$
$$2e^- + 2Fe^{3+}(aq) \rightarrow 2Fe^{2+}(aq)$$

Add $$2Fe^{3+}(aq) + H_2S(g) \rightarrow S(s) + 2H^+(aq) + 2Fe^{2+}(aq)$$

Hydrogen sulphide reduces chlorine to hydrogen chloride.

$$H_2S(g) + Cl_2(g) \rightarrow 2HCl(g) + S(s)$$
hydrogen sulphide + chlorine → hydrogen chloride + sulphur

(iv) Hydrogen sulphide as a weak dibasic acid
Hydrogen sulphide solution is feebly acidic due to the following ionisation.

$$H_2S \rightleftharpoons 2H^+ + S^{2-}$$

Hydrogen sulphide reacts with alkalis to form salts called sulphides.
E.g. Hydrogen sulphide reacts with sodium hydroxide to form sodium sulphide.

$$2NaOH(aq) + H_2S(aq) \rightarrow Na_2S(aq) + 2H_2O(l)$$
sodium hydroxide + hydrogen sulphide → sodium sulphide + water

---

24 Oxides of sulphur

24.1 INTRODUCTION

There are two oxides of sulphur—sulphur dioxide SO_2
sulphur trioxide-sulphur(VI) oxide SO_3

Sulphur dioxide is more frequently obtained. There are some similarities between sulphur dioxide and carbon dioxide.

Sulphur dioxide is made in large quantities industrially and then converted to sulphur trioxide and finally sulphuric acid (see 25.2).

24.2 PREPARATION OF SULPHUR DIOXIDE

(*i*) Sulphur dioxide can be produced by burning sulphur in air or oxygen.

$$S(s) + O_2(g) \rightarrow SO_2(g)$$
sulphur + oxygen → sulphur dioxide

This reaction is not suitable for preparing pure samples of sulphur dioxide in the laboratory.

(*ii*) Sulphur dioxide is usually prepared by the action of hot, concentrated sulphuric acid on copper turnings. (Dilute sulphuric acid does not react with copper.)

$$Cu(s) + 2H_2SO_4(l) \rightarrow CuSO_4(aq) + 2H_2O(l) + SO_2(g)$$
copper + sulphuric acid → copper sulphate + water + sulphur dioxide

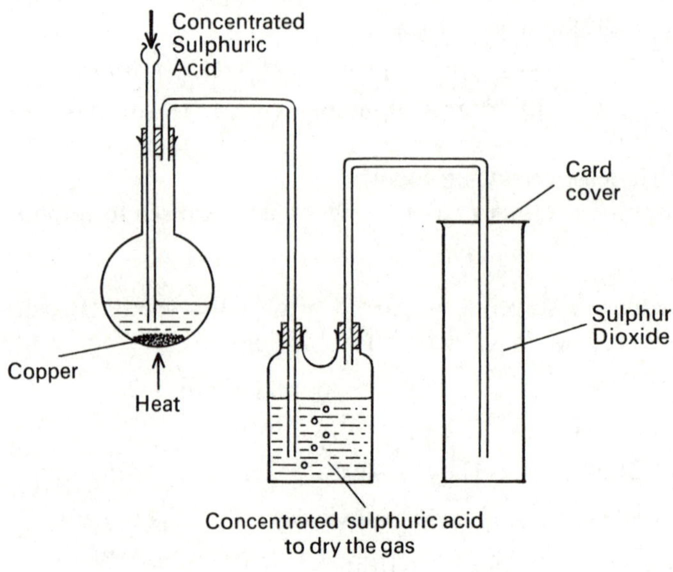

Fig. 24.1 Preparation of sulphur dioxide

The apparatus used is shown in Fig. 24.1 The sulphur dioxide is dried by passing it through concentrated sulphuric acid and collected by downward delivery because it is denser than air.

(A common error seen on examination papers is the following equation:

$$Cu(s) + H_2SO_4(l) \rightarrow CuO(s) + H_2O(l) + SO_2(g)$$

Clearly, copper(II) oxide cannot exist in the presence of excess concentrated sulphuric acid.)

(*iii*) Sulphur dioxide is prepared by the action of hot, dilute hydrochloric acid on sodium sulphite crystals.

$$Na_2SO_3(s) + 2HCl(aq) \rightarrow 2NaCl(aq) + SO_2(g) + H_2O(l)$$

sodium sulphite + hydrochloric acid → sodium chloride + sulphur dioxide + water

This is analogous to the preparation of carbon dioxide but heat is required (see 34.1).

24.3 Tests for sulphur dioxide

A piece of filter paper, soaked in orange potassium dichromate solution, turns green in the presence of sulphur dioxide. Similarly, a piece of filter paper soaked in purple potassium permanganate solution turns colourless. In both of these tests the sulphur dioxide reduces the coloured substance.

Hydrogen sulphide also affects potassium dichromate and potassium permanganate solutions in a similar way. However, hydrogen sulphide forms a precipitate of sulphur in these reactions which will blacken a piece of filter paper soaked in lead nitrate solution. Sulphur dioxide affects potassium dichromate and potassium permanganate solutions but does not blacken a piece of filter paper soaked in lead nitrate solution.

Sulphur dioxide, like carbon dioxide, turns limewater milky but there is no confusion in practice because of the pungent smell of sulphur dioxide.

24.4 Properties of sulphur dioxide

Sulphur is a colourless gas with a distinctive choking smell.

It dissolves in water to form the weak acid, sulphurous acid. Sulphur dioxide is the anhydride of sulphurous acid

$$H_2O(l) + SO_2(g) \rightleftharpoons H_2SO_3(aq)$$

water + sulphur dioxide ⇌ sulphurous acid

Carbon dioxide dissolves in a similar way to form carbonic acid H_2CO_3.

24.5 Sulphur dioxide as a reducing agent

Sulphur dioxide, in the presence of water, produces sulphurous acid, H_2SO_3, which is easily oxidised to sulphuric acid H_2SO_4. Since it is easily oxidised, it is a good reducing agent. It can only act as a reducing agent in the presence of water.

Examples

(*i*) Sulphur dioxide is a good bleaching agent because it reduces coloured dyes to colourless products.

$$\text{Dye} + H_2SO_3 \rightarrow \text{Colourless reduction product} + H_2SO_4$$

(N.B. Chlorine is also a bleaching agent but it bleaches by oxidation, see 32.9.)

(*ii*) When sulphur dioxide is passed through potassium dichromate solution, the solution turns from orange to green. The green solution contains chromium(III) sulphate.

$$K_2Cr_2O_7(aq) + 3SO_2(g) + H_2SO_4(aq) \rightarrow K_2SO_4(aq) + Cr_2(SO_4)_3(aq) + H_2O(l)$$

potassium dichromate + sulphur dioxide + sulphuric acid → potassium sulphate + chromium(III) sulphate + water

(*iii*) With chlorine (see 16.6).

24.6 Sulphur dioxide as an oxidising agent

Sulphur dioxide can sometimes act as an oxidising agent, and when it does, sulphur is always produced.

Examples

(*i*) When a piece of strongly burning magnesium is put into a gas jar containing sulphur dioxide, the magnesium continues to burn.

$$2Mg(s) + SO_2(g) \rightarrow 2MgO(s) + S(s)$$
$$\text{magnesium} + \text{sulphur dioxide} \rightarrow \text{magnesium oxide} + \text{sulphur}$$

(A similar reaction takes place with carbon dioxide in place of sulphur dioxide, see 34.2.)

(*ii*) With hydrogen sulphide (see 16.6).

24.7 Uses of sulphur dioxide

Most of the sulphur dioxide produced is used to make sulphuric acid. It is also used for bleaching wood pulp and as a preservative in foods.

24.8 Preparation of sulphur trioxide

Sulphur trioxide is not produced directly by burning sulphur in air or oxygen.

Examples

(*i*) Catalytic oxidation of sulphur dioxide.

$$2SO_2(g) + O_2(g) \rightleftharpoons 2SO_3(g)$$
$$\text{sulphur dioxide} + \text{oxygen} \rightleftharpoons \text{sulphur trioxide}$$

Pure sulphur dioxide and oxygen are mixed and passed over a heated catalyst. The catalyst can be vanadium(V) oxide (vanadium pentoxide) or platinised ceramic wool.

When the gases are cooled in an ice mixture, the sulphur trioxide sublimes and forms colourless, needle-shaped crystals. The apparatus is shown in Fig. 24.2. The anhydrous calcium chloride tube is to prevent water vapour entering the apparatus.

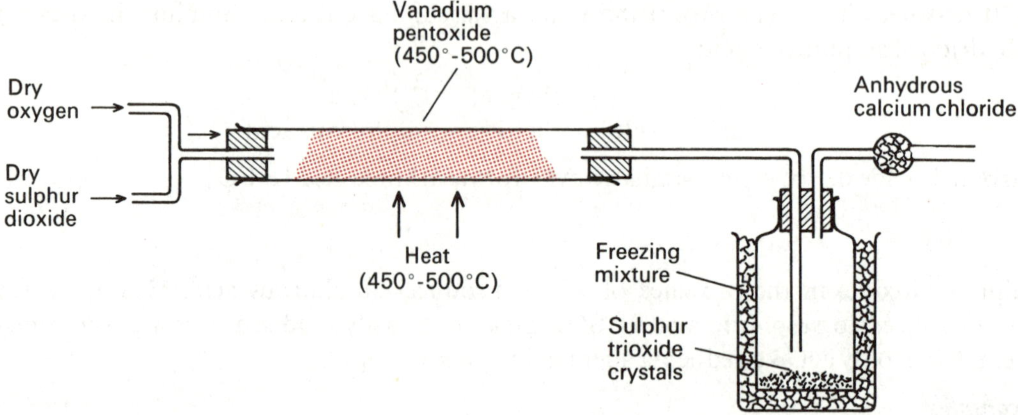

Fig. 24.2 Preparation of sulphur trioxide

(*ii*) Action of heat on certain metal sulphates (see 25.9).

24.9 Properties of sulphur trioxide

Sulphur trioxide is the anhydride of sulphuric acid.

$$SO_3(g) + H_2O(l) \rightarrow H_2SO_4(aq)$$
$$\text{sulphur trioxide} + \text{water} \rightarrow \text{sulphuric acid}$$

As usual, the oxide containing the greater percentage of oxygen (*i.e.* SO_3) is more strongly acidic.

25 Sulphuric acid and sulphates

25.1 INTRODUCTION

Sulphuric acid (sulphuric(VI) acid) is regarded as the most important product of the chemical industry. It is produced, in very large quantities, by the Contact process starting from sulphur or sulphur-containing minerals.

25.2 INDUSTRIAL MANUFACTURE OF SULPHURIC ACID (CONTACT PROCESS)

(i) Production of sulphur dioxide

Sulphur dioxide is produced by burning sulphur or heating other sulphide minerals in air.

E.g.

$$S(s) + O_2(g) \rightarrow SO_2(g)$$

sulphur + oxygen → sulphur dioxide

$$2ZnS(s) + 3O_2(g) \rightarrow 2ZnO(s) + 2SO_2(g)$$

zinc sulphide + oxygen → zinc oxide + sulphur dioxide

$$4FeS_2(s) + 11O_2(g) \rightarrow 2Fe_2O_3(s) + 8SO_2(g)$$

iron pyrites + oxygen → iron(III) oxide + sulphur dioxide

Sulphur dioxide is also produced by oxidation of hydrogen sulphide found in certain natural gas samples.

$$2H_2S(g) + 3O_2(g) \rightarrow 2H_2O(g) + 2SO_2(g)$$

hydrogen sulphide + oxygen → water + sulphur dioxide

Anhydrite (anhydrous calcium sulphate $CaSO_4$) is a useful source of sulphur dioxide in the United Kingdom. The anhydrite is finely ground with coke, sand and aluminium oxide. The mixture is strongly heated and anhydrite is reduced by the coke.

$$2CaSO_4(s) + C(s) \rightarrow 2CaO(s) + 2SO_2(g) + CO_2(g)$$

calcium sulphate + carbon → calcium oxide + sulphur dioxide + carbon dioxide

(ii) Purification of the sulphur dioxide

The sulphur dioxide produced by these methods contains certain impurities, *e.g.* arsenic compounds which, if not removed, would prevent the catalyst working ('poison the catalyst') in the next step.

The sulphur dioxide is passed through the electrostatic dust precipitators which remove charged particles, *e.g.* dust. The sulphur dioxide is then washed with water and dried.

(iii) Catalytic oxidation of sulphur dioxide (see 24.8)

$$2SO_2(g) + O_2(g) \rightleftharpoons 2SO_3(g) \qquad (\Delta H \text{ is negative})$$

sulphur dioxide + oxygen ⇌ sulphur trioxide

This is the vital step in the production of sulphuric acid. The final yield of sulphuric acid is going to depend upon how efficiently this step can be carried out.

The mixture of sulphur dioxide and air (containing oxygen) is passed through the catalyst chamber containing pellets of vanadium (V) oxide (vanadium pentoxide) V_2O_5 heated to about 450°C.

In this step, decreasing the temperature increases the yield of sulphur trioxide (and therefore of sulphuric acid) but slows down the speed of production. A compromise is made in order to enable as much sulphur trioxide to be made as possible in a reasonable time.

(iv) Absorption of sulphur trioxide

In theory, if sulphur trioxide is dissolved in water, sulphuric acid is produced but this is not done in practice on a large scale because the reaction is too exothermic and boils the sulphuric acid produced.

The sulphur trioxide is dissolved first in concentrated sulphuric acid to form oleum (fuming sulphuric acid).

$$SO_3(g) + H_2SO_4(l) \rightarrow H_2S_2O_7(l)$$
sulphur trioxide + sulphuric acid → oleum

The oleum is then diluted with the correct amount of water to produce concentrated sulphuric acid.

$$H_2S_2O_7(l) + H_2O(l) \rightarrow 2H_2SO_4(l)$$
oleum + water → sulphuric acid

The acid produced is very pure because all the impurities have been removed during the production.

25.3 Properties of sulphuric acid

Concentrated sulphuric acid is a colourless oily liquid, which does not show any acidic properties unless water is present. A great deal of heat is produced when concentrated sulphuric acid is diluted with water. It is therefore sensible to add the acid to water (rather than water to acid).

Concentrated sulphuric acid is hygroscopic. For this reason it is a good drying agent for most gases (*e.g.* SO_2).

The properties of sulphuric acid can be remembered under four headings:
(*i*) as an acid—when dilute;
(*ii*) as a producer of other acids—when concentrated;
(*iii*) as a dehydrating agent—when concentrated;
(*iv*) as an oxidising agent—when concentrated.

25.4 Sulphuric acid as an acid

Sulphuric acid is a dibasic mineral acid. It contains two replaceable hydrogens per molecule.

In the presence of water, sulphuric acid shows the usual acidic properties (see 19.2).

Dilute sulphuric acid turns blue litmus red.

Dilute sulphuric acid produces hydrogen gas with magnesium or zinc.

$$Mg(s) + H_2SO_4(aq) \rightarrow MgSO_4(aq) + H_2(g)$$
magnesium + sulphuric acid → magnesium sulphate + hydrogen

Dilute sulphuric acid produces carbon dioxide gas with a metal carbonate or hydrogen-carbonate, *e.g.* sodium carbonate.

$$Na_2CO_3(s) + H_2SO_4(aq) \rightarrow Na_2SO_4(aq) + H_2O(l) + CO_2(g)$$
sodium carbonate + sulphuric acid → sodium sulphate + water + carbon dioxide

Dilute sulphuric acid produces salts (called sulphates) with metal oxides and hydroxides.
E.g.
$$CuO(s) + H_2SO_4(aq) \rightarrow CuSO_4(aq) + H_2O(l)$$
copper(II) oxide + sulphuric acid → copper(II) sulphate + water

Because there are two replaceable hydrogens, it is possible to form acid salts.
E.g.
$$2NaOH(aq) + H_2SO_4(aq) \rightarrow Na_2SO_4(aq) + 2H_2O(l)$$
sodium hydroxide + sulphuric acid → sodium sulphate + water

$$NaOH(aq) + H_2SO_4(aq) \rightarrow NaHSO_4(aq) + H_2O(l)$$
sodium hydroxide + sulphuric acid → sodium hydrogensulphate + water

Dilute sulphuric acid can be distinguished from other dilute acids because it is a sulphate and gives a positive sulphate test (see 41.2).

25.5 SULPHURIC ACID AS A PRODUCER OF OTHER ACIDS

Concentrated sulphuric acid can be used to prepare nitric and hydrochloric acids.

If concentrated sulphuric acid is added to any metal nitrate (*e.g.* sodium nitrate) and the mixture is heated, nitric acid vapour is produced (see 28.2).

$$NaNO_3(s) + H_2SO_4(l) \rightarrow NaHSO_4(s) + HNO_3(g)$$
sodium nitrate + sulphuric acid → sodium hydrogensulphate + nitric acid

If concentrated sulphuric acid is added to any metal chloride (*e.g.* sodium chloride), hydrogen chloride gas is produced, which dissolves in water to form hydrochloric acid. Gentle heating may be necessary (see 32.1).

$$NaCl(s) + H_2SO_4(l) \rightarrow NaHSO_4(s) + HCl(g)$$
sodium chloride + sulphuric acid → sodium hydrogensulphate + hydrogen chloride

Both of these reactions take place because the nitric acid and hydrogen chloride have lower boiling points than sulphuric acid. (They are more volatile.) They escape from the reaction mixture in preference to sulphuric acid.

25.6 SULPHURIC ACID AS A DEHYDRATING AGENT

This is a consequence of the affinity of concentrated sulphuric acid for water.

(i) Sugar $C_{12}H_{22}O_{11}$
Sugar (sucrose) is a carbohydrate and contains the elements of water (hydrogen and oxygen).

If concentrated sulphuric acid is added to a sample of sugar, the sugar turns yellow, then brown, and finally black. The black solid residue is carbon which is formed when the concentrated sulphuric acid has removed the hydrogen and oxygen. The reaction is very exothermic.

$$C_{12}H_{22}O_{11}(s) \rightarrow 12C(s) + 11H_2O(g)$$
sugar → carbon + water

Similar reactions take place when other carbohydrates are used. For this reason concentrated sulphuric acid has to be used carefully with carbon compounds.

(ii) Copper(II) sulphate crystals $CuSO_4 . 5H_2O$
Copper(II) sulphate crystals contain water of crystallisation. When concentrated sulphuric acid is added to blue copper(II) sulphate crystals, the crystals turn white because the water of crystallisation has been removed by the concentrated sulphuric acid.

$$CuSO_4.5H_2O(s) \rightleftharpoons CuSO_4(s) + 5H_2O(l)$$
copper sulphate crystals ⇌ anhydrous copper(II) sulphate + water

(iii) Methanoic (formic) acid and ethanedioic (oxalic) acid
Both of these acids are dehydrated by concentrated sulphuric acid.

$$HCOOH(s) \rightarrow H_2O(l) + CO(g)$$
methanoic (formic) acid → water + carbon monoxide
$$H_2C_2O_4(s) \rightarrow H_2O(l) + CO(g) + CO_2(g)$$
ethanedioic (oxalic) acid → water + carbon monoxide + carbon dioxide

25.7 SULPHURIC ACID AS AN OXIDISING AGENT

Concentrated sulphuric acid acts as an oxidising agent in a wide range of reactions. Usually the sulphuric acid is hot and concentrated. In each case the sulphuric acid (H_2SO_4) is reduced to water and sulphur dioxide.

(i) Copper

$$Cu(s) + 2H_2SO_4(l) \rightarrow CuSO_4(aq) + 2H_2O(l) + SO_2(g)$$
copper + sulphuric acid → copper(II) sulphate + water + sulphur dioxide

(ii) Carbon and sulphur

Hot concentrated sulphuric acid oxidises carbon and sulphur to carbon dioxide and sulphur dioxide respectively.

$$C(s) + 2H_2SO_4(l) \rightarrow CO_2(g) + 2SO_2(g) + 2H_2O(g)$$

carbon + sulphuric acid → carbon dioxide + sulphur dioxide + water

$$S(s) + 2H_2SO_4(l) \rightarrow 3SO_2(g) + 2H_2O(g)$$

sulphur + sulphuric acid → sulphur dioxide + water

25.8 Uses of sulphuric acid

The uses of sulphuric acid include:

(*i*) manufacture of fertilisers, *e.g.* ammonium sulphate;
(*ii*) making drugs, dyes and other chemicals;
(*iii*) making synthetic detergents;
(*iv*) removing the surface oxide ('pickling') from metals;
(*v*) making rayon.

25.9 Sulphates

Sulphates are salts produced using sulphuric acid. All common sulphates are soluble in water except barium sulphate $BaSO_4$ and lead sulphate $PbSO_4$. (Calcium sulphate $CaSO_4$ is only slightly soluble.)

Sulphates are less easily decomposed by heating than nitrates or carbonates. Iron(II) sulphate and iron(III) sulphate decompose on heating.

$$2FeSO_4.7H_2O(s) \rightarrow Fe_2O_3(s) + SO_2(g) + SO_3(g) + 14H_2O(g)$$

iron(II) sulphate → iron(III) oxide + sulphur dioxide + sulphur trioxide + water

$$Fe_2(SO_4)_3(s) \rightarrow Fe_2O_3(s) + 3SO_3(g)$$

iron(III) sulphate → iron(III) oxide + sulphur trioxide

Test for a sulphate (see 41.2).

26 Nitrogen

26.1 Introduction

Nitrogen is the major constituent of air but it is very unreactive. This is due to the very strong triple covalent bond between the pair of nitrogen atoms in a nitrogen (N_2) molecule. It is regarded as the 'inactive' gas which dilutes the 'active' gas, oxygen, in air.

26.2 Preparation of nitrogen from air

When water is run into the reservoir of air in the apparatus in Fig. 26.1, air is pushed through the apparatus.

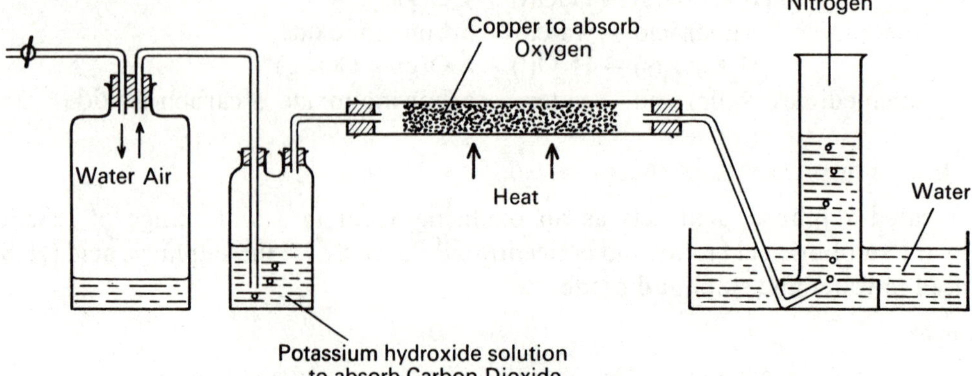

Fig. 26.1 Preparation of nitrogen from air

When the air is bubbled through potassium hydroxide solution, the carbon dioxide is removed.

$$CO_2(g) + 2KOH(aq) \rightarrow K_2CO_3(aq) + H_2O(l)$$
carbon dioxide + potassium hydroxide → potassium carbonate + water

Oxygen is removed from the air when the air is passed over heated copper.

$$2Cu(s) + O_2(g) \rightarrow 2CuO(s)$$
copper + oxygen → copper(II) oxide

The remaining gas is collected over water. It is not, however, pure nitrogen because it contains the noble or inert gases (*e.g.* argon). It has the same chemical properties as nitrogen but its density is slightly different.

26.3 Laboratory preparation of pure nitrogen

Pure nitrogen is produced when ammonium nitrite solution is heated using the apparatus in Fig. 26.2.

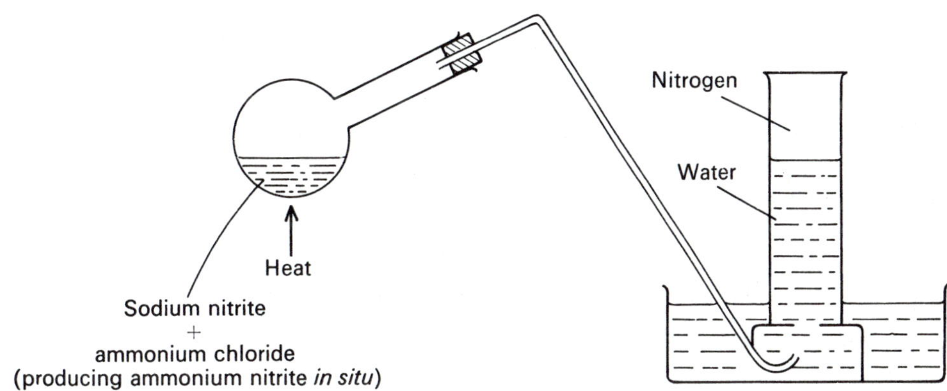

Fig. 26.2 Preparation of nitrogen from ammonium nitrite

$$NH_4NO_2(s) \rightarrow N_2(g) + 2H_2O(g)$$
ammonium nitrite → nitrogen + water

(N.B. Heating ammonium nitrate produces dinitrogen monoxide N_2O.)

Ammonium nitrite is prepared as it is required (*in situ*) by mixing together concentrated solutions of ammonium chloride and sodium nitrite. This is because ammonium nitrite is too unstable to store.

The nitrogen gas is collected over water. If it is required dry, it is dried by passing it through concentrated sulphuric acid and then it must be collected over mercury.

26.4 Industrial manufacture of nitrogen

Fractional distillation of liquid air (see 8.2).

26.5 Compounds that produce nitrogen on heating

Apart from ammonium nitrite, other compounds produce nitrogen when heated. Ammonium dichromate $(NH_4)_2Cr_2O_7$ decomposes on heating to produce nitrogen but the reaction is too exothermic to be used for the laboratory preparation.

$$(NH_4)_2Cr_2O_7(s) \rightarrow N_2(g) + Cr_2O_3(s) + 4H_2O(g)$$
ammonium dichromate → nitrogen + chromium(III) oxide + water

Ammonia gas decomposes when passed over strongly heated iron wool.

$$2NH_3(g) \rightarrow N_2(g) + 3H_2(g)$$
ammonia → nitrogen + hydrogen

This is not suitable for a laboratory preparation because a mixture of gases is produced. Nitrogen can be produced by passing this mixture of gases over heated copper(II) oxide, which removes the hydrogen.

26.6 Test for nitrogen

It is not easy to test for nitrogen because of its unreactivity. If an odourless gas sample extinguishes a lighted splint and does not turn limewater milky, it could well be nitrogen.

The only positive test for nitrogen is to burn magnesium in it to form magnesium nitride. When water is added to magnesium nitride, ammonia gas is evolved.

$$Mg_3N_2(s) + 6H_2O(l) \rightarrow 3Mg(OH)_2(s) + 2NH_3(g)$$
magnesium nitride + water → magnesium hydroxide + ammonia

26.7 Properties of nitrogen

Nitrogen is a colourless, odourless and tasteless gas. It is only very slightly soluble in water and its density is about the same as air.

It is very unreactive and reacts only with the reactive metals (magnesium, calcium and lithium). When the metals are heated strongly, they burn in nitrogen forming the corresponding nitride. These nitrides are pale yellow ionic compounds containing the N^{3-} ion.

E.g.

$$3Mg(s) + N_2(g) \rightarrow Mg_3N_2(s)$$
magnesium + nitrogen → magnesium nitride

The only other reactions of nitrogen occur under conditions too drastic for the laboratory.

(i) Hydrogen
When a mixture of nitrogen and hydrogen are passed over a strongly heated iron catalyst, ammonia is produced (Haber process, see 27.3).

$$N_2(g) + 3H_2(g) \rightleftharpoons 2NH_3(g)$$
nitrogen + hydrogen ⇌ ammonia

(ii) Oxygen
When a mixture of nitrogen and oxygen are subjected to the spark from an electric arc (or lightning), the gases combine to form nitrogen monoxide.

$$N_2(g) + O_2(g) \rightarrow 2NO(g)$$
nitrogen + oxygen → nitrogen monoxide

26.8 Uses of nitrogen

Most of the nitrogen used in industry is used to produce ammonia (see 27.3) and ammonia is used to make nitric acid (see 28.3).

26.9 Nitrogen cycle

Nitrogen is an essential element for plant and animal life. It is a constituent of proteins that are necessary in all plant and animal cells. Despite the large amount of nitrogen in the air, this cannot be used directly by animals (including man) and so animals get their nitrogen from plant proteins. Most plants obtain their nitrogen in the form of nitrates through their roots.

Fig. 26.3 shows the circulation of nitrogen in nature—called **the nitrogen cycle**.

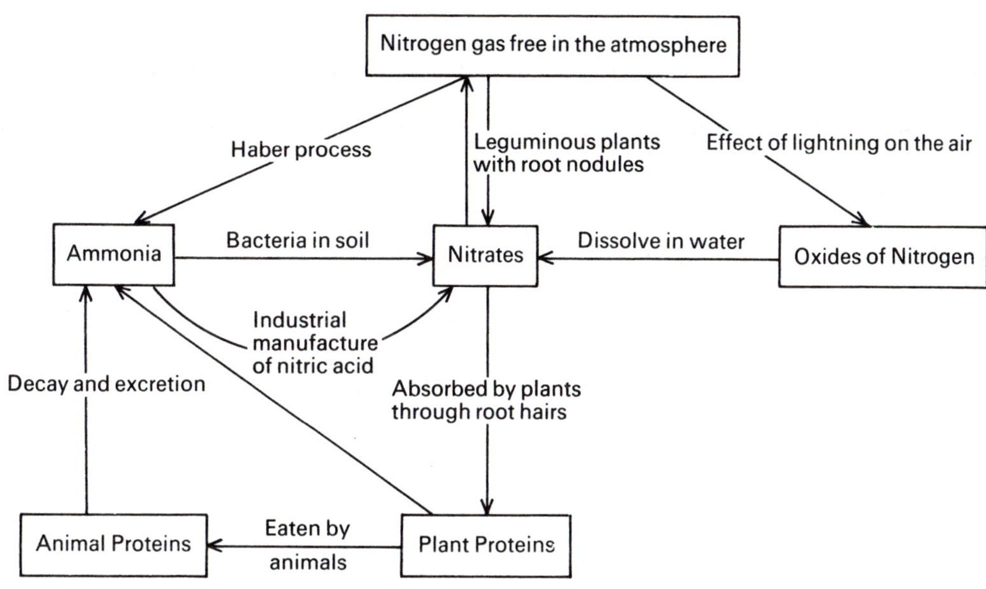

Fig. 26.3 The nitrogen cycle

Nitrogen in the air is fixed in the soil by three methods:

(*i*) Lightning causes nitrogen and oxygen to react together forming nitrogen monoxide (see 26.7). This nitrogen monoxide finally forms nitrates in the soil.

(*ii*) Bacteria in root nodules of certain plants (called leguminous plants, *e.g.* clover, peas *etc.*) are able to absorb nitrogen directly from the air.

(*iii*) Certain bacteria in the soil are able to fix nitrogen directly from the air.

Nitrogen also enters the soil from the death and decay of plants and animals, from animal urine and faeces. Bacterial action converts proteins into ammonia and then, via nitrites, into nitrates.

Because man intervenes in the nitrogen cycle by removing crops from the soil and not allowing them to decay, it becomes necessary to add nitrogen to the soil in the form of fertilisers (see 30.2).

27 Ammonia and ammonium compounds

27.1 Introduction

Ammonia (NH_3) is a compound of nitrogen and hydrogen. It is the only common gas that turns damp red litmus blue, *i.e.* it is alkaline. It reacts with acids to form solid salts called ammonium compounds. Ammonium compounds contain the ammonium ion (NH_4^+).

27.2 Laboratory preparation of ammonia

Ammonia gas is prepared in the laboratory by heating a mixture of an ammonium salt and an alkali. For example, a mixture of ammonium chloride and sodium hydroxide could be used.

$$NH_4Cl(s) + NaOH(s) \rightarrow NaCl(s) + H_2O(g) + NH_3(g)$$
ammonium chloride + sodium hydroxide → sodium chloride + water + ammonia

or

$$(NH_4)_2SO_4(s) + Ca(OH)_2(s) \rightarrow CaSO_4(s) + 2H_2O(g) + 2NH_3(g)$$
ammonium sulphate + calcium hydroxide → calcium sulphate + water + ammonia

The basic reaction taking place in each case can be represented by the same ionic equation:

$$NH_4^+ + OH^- \rightarrow NH_3 + H_2O$$

Suitable apparatus for preparing dry ammonia is shown in Fig. 27.1.

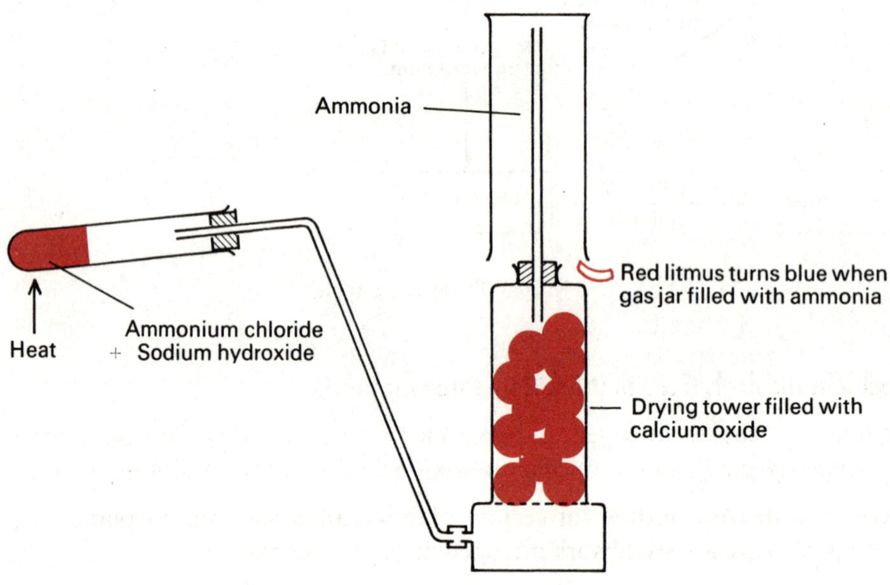

Fig. 27.1 Preparation of ammonia

Ammonia gas is dried by passing it through a tower containing calcium oxide (quicklime).

$$CaO(s) + H_2O(l) \rightarrow Ca(OH)_2(s)$$
calcium oxide + water → calcium hydroxide

(Concentrated sulphuric acid cannot be used because it reacts with ammonia gas

$$2NH_3(g) + H_2SO_4(l) \rightarrow (NH_4)_2SO_4(s)$$
ammonia + sulphuric acid → ammonium sulphate)

Ammonia gas is collected by upward delivery (downward displacement of air) because it is much less dense than air and is readily soluble in water.

27.3 Industrial production of ammonia

Ammonia is produced industrially in large amounts from nitrogen and hydrogen using the **Haber process**.

(i) Sources of nitrogen and hydrogen

Nitrogen is obtained by fractional distillation of liquid air (see 8.2). Hydrogen can be obtained from water (see 10.4) and also from oil or natural gas (see 10.4).

(ii) Reaction between nitrogen and hydrogen

The mixture of nitrogen (1 part) and hydrogen (3 parts) is compressed to 200 atmospheres and passed through the catalyst chamber at 500°C. The catalyst chamber contains finely divided iron as the catalyst. About 10% of the nitrogen and hydrogen mixture is converted to ammonia.

$$N_2(g) + 3H_2(g) \rightleftharpoons 2NH_3(g)$$
nitrogen + hydrogen ⇌ ammonia (Forward reaction exothermic, *i.e.* ΔH is negative.)

If the temperature is decreased, there is a greater conversion of nitrogen and hydrogen to ammonia but the reaction is much slower.

(iii) Removal of ammonia from the mixture of gases

When the mixture of gases leaving the catalyst chamber is cooled, only ammonia liquefies and can be removed. The unreacted nitrogen and hydrogen are re-cycled.

27.4 Testing for ammonia

(*i*) Ammonia turns red litmus blue and does not burn in air when a lighted splint is applied.

(*ii*) Dense white fumes of ammonium chloride are formed when ammonia gas comes into contact with hydrogen chloride gas (*e.g.* the stopper from a bottle of concentrated hydrochloric acid).

$$NH_3(g) + HCl(g) \rightleftharpoons NH_4Cl(s)$$

27.5 Properties of ammonia

Ammonia is a colourless gas with a pungent and characteristic odour (smelling salts). It turns red litmus blue and is less dense than air.

(i) Solubility in water

Ammonia dissolves very readily in cold water to form ammonia solution $NH_3(aq)$. This is sometimes called ammonium hydroxide NH_4OH. This solution is alkaline.

$$NH_3(g) + H_2O(l) \rightleftharpoons NH_4^+(aq) + OH^-(aq)$$

The solubility of ammonia in water can be demonstrated by the fountain experiment. A dry flask is filled with dry ammonia and set up as shown in Fig. 27.2. The end of the tube is dipped into water. The water level rises in the tube as the ammonia dissolves and the water fountains into the flask.

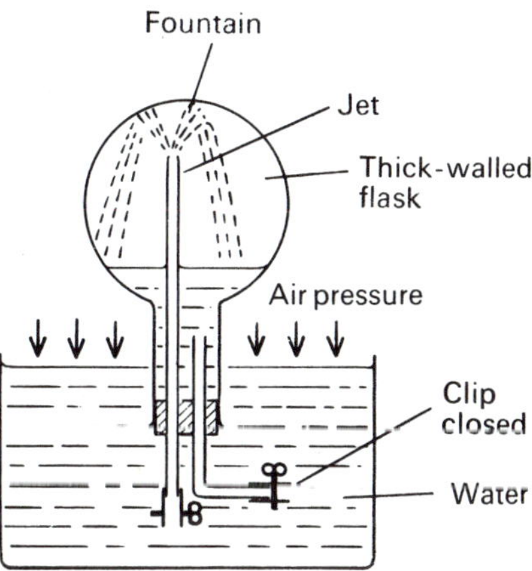

Fig. 27.2 Fountain experiment

A similar experiment can be used to demonstrate the solubility of hydrogen chloride in water. The apparatus required to produce a solution of ammonia in water from ammonia gas is shown in Fig. 32.2.

(ii) Reactions with acids

Ammonia gas is absorbed by acids to form ammonium salts.

E.g. $NH_3(g) + HCl(aq) \rightarrow NH_4Cl(aq)$
ammonia + hydrochloric acid → ammonium chloride

$NH_3(g) + HNO_3(aq) \rightarrow NH_4NO_3(aq)$
ammonia + nitric acid → ammonium nitrate

$2NH_3(g) + H_2SO_4(aq) \rightarrow (NH_4)_2SO_4(aq)$
ammonia + sulphuric acid → ammonium sulphate

(iii) Combustion of ammonia

Ammonia does not burn in air but it burns in oxygen (or air enriched with oxygen).

$4NH_3(g) + 3O_2(g) \rightarrow 2N_2(g) + 6H_2O(g)$
ammonia + oxygen → nitrogen + water

Substances do not burn in ammonia gas.

It is possible to oxidise ammonia with oxygen in the presence of a heated platinum catalyst to form nitrogen monoxide. A heated platinum wire glows brightly in a mixture of ammonia and oxygen because this reaction is very exothermic.

$4NH_3(g) + 5O_2(g) \rightarrow 4NO(g) + 6H_2O(g)$
ammonia + oxygen → nitrogen monoxide + water

This is the basis of the industrial manufacture of nitric acid from ammonia (see 28.3).

(iv) Ammonia as a reducing agent

When ammonia gas is passed over heated copper(II) oxide, the ammonia is oxidised to nitrogen and the copper(II) oxide is reduced to copper (Fig. 27.3).

$2NH_3(g) + 3CuO(s) \rightarrow 3Cu(s) + 3H_2O(g) + N_2(g)$
ammonia + copper(II) oxide → copper + water + nitrogen
(black) (pinkish-brown)

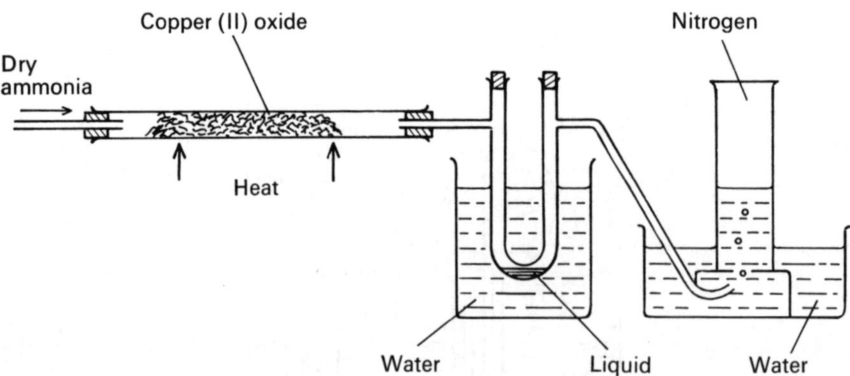

Fig. 27.3 Production of nitrogen by the passing of ammonia over copper(II) oxide

A similar reaction takes place with oxides of lead and iron. This experiment can be used to demonstrate that ammonia contains nitrogen.

When chlorine and ammonia gases are mixed, chlorine is reduced to hydrogen chloride by ammonia.

$$2NH_3(g) + 3Cl_2(g) \rightarrow N_2(g) + 6HCl(g)$$
ammonia + chlorine → nitrogen + hydrogen chloride

However, a further reaction takes place

$$6NH_3(g) + 6HCl(g) \rightarrow 6NH_4Cl(s)$$
ammonia + hydrogen chloride → ammonium chloride

Adding these two equations together gives the equation for the overall reaction.

$$8NH_3(g) + 3Cl_2(g) \rightarrow N_2(g) + 6NH_4Cl(s)$$
ammonia + chlorine → nitrogen + ammonium chloride

27.6 Uses of ammonia

(*i*) Ammonia is used to manufacture fertilisers, *e.g.* ammonium sulphate.
(*ii*) Ammonia is used to manufacture nitric acid.
(*iii*) Ammonia is used to manufacture plastics.

27.7 Ammonium compounds

Ammonium compounds contain the ammonium ion (NH_4^+). This ion is produced when a hydrogen ion H^+ (or proton) is bonded to the spare pair of electrons on the nitrogen atom in the ammonia molecule, forming a co-ordinate bond (see 5.3). The ammonium ion formed is tetrahedral.

Although ammonium compounds contain no metal ions, they are very similar to the corresponding alkali metal compounds in many ways. Ammonium chloride and sodium chloride are both colourless, crystalline solids that dissolve in water.

Ammonia solution (ammonium hydroxide) is an important alkali and is a source of all ammonium compounds.

It precipitates metal hydroxides from solutions of certain metal salts (see 41.3).

E.g. $$FeCl_2(aq) + 2NH_4OH(aq) \rightarrow Fe(OH)_2(s) + 2NH_4Cl(aq)$$
iron(II) chloride + ammonium hydroxide → iron(II) hydroxide + ammonium chloride

$$FeCl_3(aq) + 3NH_4OH(aq) \rightarrow Fe(OH)_3(s) + 3NH_4Cl(aq)$$
iron(III) chloride + ammonium hydroxide → iron(III) hydroxide + ammonium chloride

$$CuSO_4(aq) + 2NH_4OH(aq) \rightarrow Cu(OH)_2(s) + (NH_4)_2SO_4$$
copper(II) sulphate + ammonium hydroxide → copper(II) hydroxide
+ ammonium sulphate

In the case of copper(II) hydroxide precipitation, the copper (II) hydroxide precipitate redissolves and forms a deep blue solution containing the complex tetraammine copper(II) ion.

The following general remarks about ammonium compounds are worth remembering:

(*i*) Most ammonium salts are white, crystalline solids (exception—ammonium dichromate is orange).

(*ii*) All ammonium salts are soluble in water.

(*iii*) All ammonium salts liberate ammonia gas when heated with an alkali (see 27.2).

(*iv*) Ammonium compounds are not very stable thermally. They sublime or decompose on heating. Ammonium chloride and ammonium carbonate sublime. Ammonium nitrate (see 28.6), ammonium nitrite (see 26.3) and ammonium dichromate (see 26.5) decompose on heating.

28 Nitric acid and nitrates

28.1 Introduction

Nitric acid, sometimes called nitric(V) acid, is a common monobasic mineral acid. Replacement of the hydrogen ion in the nitric acid by a metal ion produces a nitrate.

E.g.

$$HNO_3 \rightarrow NaNO_3$$
nitric acid → sodium nitrate

Concentrated nitric acid is a colourless, fuming liquid.

28.2 Laboratory preparation of nitric acid

In the laboratory, nitric acid is prepared by heating a mixture of potassium nitrate and concentrated sulphuric acid. The nitric acid distils over because it is more volatile.

$$KNO_3(s) + H_2SO_4(l) \rightarrow HNO_3(g) + KHSO_4(s)$$
potassium nitrate + sulphuric acid → nitric acid + potassium hydrogensulphate

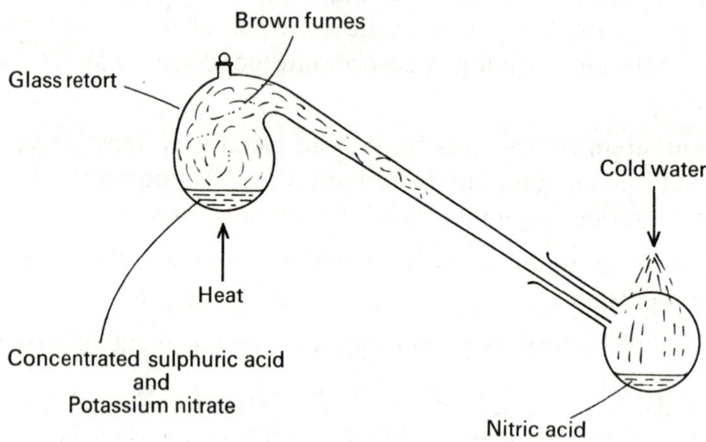

Fig. 28.1 Preparation of nitric acid

The experiment must be carried out in an all-glass apparatus (Fig. 28.1) because nitric acid vapour attacks rubber and cork. The cold water condenses the nitric acid vapour. The nitric acid produced should be colourless but it is often yellow in colour due to partial decomposition. The yellow colour can be removed by bubbling air through the acid.

28.3 Industrial manufacture of nitric acid

Nitric acid is manufactured by the catalytic oxidation of ammonia and dissolving the products in water.

(*i*) A mixture of ammonia vapour and excess air are passed over a heated platinum gauze catalyst at 900°C. An exothermic reaction takes place producing nitrogen monoxide and steam.

$$4NH_3(g) + 5O_2(g) \rightarrow 4NO(g) + 6H_2O(g)$$
ammonia + oxygen → nitrogen monoxide + water

(*ii*) The mixture of gases is cooled and nitrogen monoxide reacts with further oxygen in the air to produce nitrogen dioxide.

$$2NO(g) + O_2(g) \rightarrow 2NO_2(g)$$
nitrogen monoxide + oxygen → nitrogen dioxide

(*iii*) The mixture of gases is then passed through a tower containing a flow of cold water. Nitric acid is produced by the reaction of nitrogen dioxide with water in the presence of oxygen.

$$4NO_2(g) + O_2(g) + 2H_2O(l) \rightarrow 4HNO_3(l)$$
nitrogen dioxide + oxygen + water → nitric acid

28.4 PROPERTIES OF CONCENTRATED NITRIC ACID

(i) Action of heat on concentrated nitric acid
Thermal decomposition of concentrated nitric acid produces a mixture of nitrogen dioxide, steam and oxygen. If the gas produced is collected over water, only oxygen is collected.

$$4HNO_3(l) \rightarrow 2H_2O(g) + 4NO_2(g) + O_2(g)$$
nitric acid → water + nitrogen dioxide + oxygen

(ii) Oxidising properties of concentrated nitric acid
Concentrated nitric acid is a strong oxidising agent (see 16.1). When it acts as an oxidising agent, it is reduced to nitrogen dioxide and water. The following are examples of this property.

Copper. Concentrated nitric acid reacts with copper to form a blue solution of copper(II) nitrate. The nitric acid is reduced to nitrogen dioxide and water. The reaction is exothermic.

$$Cu(s) + 4HNO_3(l) \rightarrow Cu(NO_3)_2(aq) + 2H_2O(l) + 2NO_2(g)$$
copper + nitric acid → copper nitrate + water + nitrogen dioxide

The reaction takes place without heating. Concentrated nitric acid reacts with all common metals except gold and platinum.

Carbon. Carbon is oxidised by warm, concentrated nitric acid producing carbon dioxide. The nitric acid is reduced to nitrogen dioxide and water.

$$C(s) + 4HNO_3(l) \rightarrow CO_2(g) + 4NO_2(g) + 2H_2O(g)$$
carbon + nitric acid → carbon dioxide + nitrogen dioxide + water

Sulphur. Sulphur is oxidised to sulphuric acid by warm nitric acid.

Sawdust. Sawdust contains compounds of carbon, hydrogen and oxygen. If concentrated nitric acid is added to sawdust, after a short delay, a violent reaction takes place producing nitrogen dioxide and water. Similar reactions take place with paper, cork and rubber.

Iron(II) sulphate. Concentrated nitric acid oxidises iron(II) sulphate, dissolved in dilute sulphuric acid, to iron(III) sulphate on warming (see 41.2).

$$6FeSO_4(aq) + 3H_2SO_4(aq) + 2HNO_3(l) \rightarrow 3Fe_2(SO_4)_3(aq) + 4H_2O(l) + 2NO(g)$$
iron(II) sulphate + sulphuric acid → iron(III) sulphate + water
+ nitric acid + nitrogen monoxide

The nitrogen monoxide forms nitrogen dioxide in contact with air.

$$2NO(g) + O_2(g) \rightarrow 2NO_2(g)$$
nitrogen monoxide + oxygen → nitrogen dioxide

28.5 PROPERTIES OF DILUTE NITRIC ACID

(i) With indicators
Dilute nitric acid turns blue litmus red.

(ii) With metals
Most dilute acids will react with some metals to produce hydrogen. Hydrogen is only pro-

duced using very dilute nitric acid. If magnesium is reacted with cold, very dilute nitric acid, hydrogen is produced.

$$Mg(s) + 2HNO_3(aq) \rightarrow Mg(NO_3)_2(aq) + H_2(g)$$
magnesium + nitric acid → magnesium nitrate + hydrogen

In other cases, dilute nitric acid is still a sufficiently strong oxidising agent to oxidise the hydrogen to water.

$$3Cu(s) + 8HNO_3(aq) \rightarrow 3Cu(NO_3)_2(aq) + 4H_2O(l) + 2NO(g)$$
copper + nitric acid → copper(II) nitrate + water + nitrogen monoxide

The nitrogen monoxide forms nitrogen dioxide in contact with air.

$$2NO(g) + O_2(g) \rightarrow 2NO_2(g)$$
nitrogen monoxide + oxygen → nitrogen dioxide

More reactive metals than copper may produce dinitrogen monoxide (N_2O) or nitrogen (N_2).

(iii) With metal oxides or hydroxides

Dilute nitric acid reacts with a metal oxide or hydroxide to produce a metal nitrate solution. The mixture has to be warmed in most cases.

E.g. $$CuO(s) + 2HNO_3(aq) \rightarrow Cu(NO_3)_2(aq) + H_2O(l)$$
copper(II) oxide + nitric acid → copper nitrate + water

$$NaOH(aq) + HNO_3(aq) \rightarrow NaNO_3(aq) + H_2O(l)$$
sodium hydroxide + nitric acid → sodium nitrate + water

(iv) With metal carbonates

Dilute nitric acid reacts with metal carbonates in a similar way to other dilute acids producing carbon dioxide and water. With dilute nitric acid the salt produced is a nitrate. It is not necessary to heat the mixture.

E.g. $$CaCO_3(s) + 2HNO_3(aq) \rightarrow Ca(NO_3)_2(aq) + H_2O(l) + CO_2(g)$$
calcium carbonate + nitric acid → calcium nitrate + water + carbon dioxide

$$Na_2CO_3(s) + 2HNO_3(aq) \rightarrow 2NaNO_3(aq) + H_2O(l) + CO_2(g)$$
sodium carbonate + nitric acid → sodium nitrate + water + carbon dioxide

28.6 NITRATES

All nitrates are soluble in water

The ease of decomposition of metal nitrates when heated is related to the position of the metal in the reactivity series (see 15.1). The lower the metal is in the series, the more readily and completely the nitrate of that metal decomposes. Table 28.1 summarises the action of heat on some nitrates.

Ammonium nitrate, NH_4NO_3, decomposes on heating to produce dinitrogen monoxide (nitrous oxide N_2O). Samples of ammonium nitrate can be explosive and it is best produced *in situ* by using a mixture of ammonium chloride and potassium nitrate.

$$NH_4NO_3(s) \rightarrow N_2O(g) + 2H_2O(g)$$
ammonium nitrate → dinitrogen monoxide + water

The apparatus required to prepare and collect a sample of dinitrogen monoxide is the same as that used to prepare and collect nitrogen shown in Fig. 26.2. However, dinitrogen monoxide is collected over hot water because it is quite soluble in cold water.

Test for a nitrate (see 41.2).

Table 28.1 Action of heat on metal nitrates

Nitrate	*Formula*	*Colour*	*Equation*	*Colour of the solid residue*
Potassium nitrate	KNO_3	white	$2KNO_3(s) \rightarrow 2KNO_2(s) + O_2(g)$ potassium nitrate → potassium nitrite + oxygen	white
Sodium nitrate	$NaNO_3$	white	$2NaNO_3(s) \rightarrow 2NaNO_2(s) + O_2(g)$ sodium nitrate → sodium nitrite + oxygen	white
Calcium nitrate	$Ca(NO_3)_2$	white	$2Ca(NO_3)_2(s) \rightarrow 2CaO(s) + 4NO_2(g) + O_2(g)$ calcium nitrate → calcium oxide + nitrogen dioxide + oxygen	white
Magnesium nitrate	$Mg(NO_3)_2$	white	$2Mg(NO_3)_2(s) \rightarrow 2MgO(s) + 4NO_2(g) + O_2(g)$ magnesium nitrate → magnesium oxide + nitrogen dioxide + oxygen	white
Zinc nitrate	$Zn(NO_3)_2$	white	$2Zn(NO_3)_2(s) \rightarrow 2ZnO(s) + 4NO_2(g) + O_2(g)$ zinc nitrate → zinc oxide + nitrogen dioxide + oxygen	yellow when hot white when cold
Iron(III) nitrate	$Fe(NO_3)_3$	brown	$4Fe(NO_3)_3(s) \rightarrow 2Fe_2O_3(s) + 12NO_2(g) + 3O_2(g)$ iron(III) nitrate → iron(III) oxide + nitrogen dioxide + oxygen	reddish-brown
Lead nitrate	$Pb(NO_3)_2$	white	$2Pb(NO_3)_2(s) \rightarrow 2PbO(s) + 4NO_2(g) + O_2(g)$ lead nitrate → lead oxide + nitrogen dioxide + oxygen	yellow
Copper(II) nitrate	$Cu(NO_3)_2$	blue	$2Cu(NO_3)_2(s) \rightarrow 2CuO(s) + 4NO_2(g) + O_2(g)$ copper(II) nitrate → copper(II) oxide + nitrogen dioxide + oxygen	black
Silver nitrate	$AgNO_3$	white	$2AgNO_3(s) \rightarrow 2Ag(s) + 2NO_2(g) + O_2(g)$ silver nitrate → silver + nitrogen dioxide + oxygen	silvery

29 Oxides of nitrogen

The three common oxides of nitrogen are compared in Table 29.1. None of these oxides can be prepared in the laboratory from nitrogen and they are all relatively unstable.

Table 29.1

Name	*Dinitrogen monoxide*
Formula	N_2O
Alternative name	Nitrous oxide
Laboratory preparation	Action of heat on ammonium nitrate. $NH_4NO_3(s) \rightarrow N_2O(g) + 2H_2O(g)$ ammonium nitrate → dinitrogen monoxide + water In practice, the ammonium nitrate is prepared *in situ* from ammonium chloride and sodium nitrate. The apparatus is similar to that in Fig. 26.2 but gas is collected over warm water
Solubility in water	Slightly soluble
Effect on litmus	No effect—neutral
Test	Re-lights a glowing splint (see 9.11) but has a sweetish smell and is more soluble in water than oxygen
Reaction with sodium hydroxide solution	No reaction
Reaction with iron(II) sulphate solution	No reaction
Reaction with burning magnesium	Magnesium decomposes gas and continues to burn $Mg(s) + N_2O(g) \rightarrow MgO(s) + N_2(g)$ magnesium + dinitrogen monoxide → magnesium oxide + nitrogen

Nitrogen monoxide	*Nitrogen dioxide*
NO	NO_2
Nitric oxide	
Action of dilute nitric acid on copper (see 28.5). Apparatus as in Fig. 10.1. Some nitrogen dioxide is produced but it dissolves in water	Action of heat on lead(II) nitrate (see 28.6) Apparatus in Fig. 29.1 below. When the mixture of gases is cooled, nitrogen dioxide liquefies and forms dinitrogen tetroxide N_2O_4. $2NO_2(g) \rightleftharpoons N_2O_4(l)$ nitrogen dioxide ⇌ dinitrogen tetroxide
Almost insoluble	Readily soluble forming a mixture of nitric and nitrous acids $2NO_2(g) + H_2O(aq) \rightarrow HNO_2(aq) + HNO_3(aq)$ nitrogen dioxide + water → nitrous acid + nitric acid
No effect—neutral	Turns litmus red—acidic
When exposed to air, nitrogen monoxide forms brown fumes $2NO(g) + O_2(g) \rightarrow 2NO_2(g)$ nitrogen monoxide + oxygen → nitrogen dioxide	Brown gas which turns blue litmus red. Smells similar to chlorine but unlike bromine has no bleaching properties
No reaction	Reacts $2NO_2(g) + 2NaOH(aq) \rightarrow NaNO_2(aq) + NaNO_3(aq) + H_2O(l)$ nitrogen dioxide + sodium hydroxide → sodium nitrite + sodium nitrate + water
Forms a brown complex (see 41.2)	No reaction
Magnesium decomposes gas and continues to burn. $2Mg(s) + 2NO(g) \rightarrow 2MgO(s) + N_2(g)$ magnesium + nitrogen monoxide → magnesium oxide + nitrogen	Magnesium decomposes gas and continues to burn. $4Mg(s) + 2NO_2(g) \rightarrow 4MgO(s) + N_2(g)$ magnesium + nitrogen dioxide → magnesium oxide + nitrogen

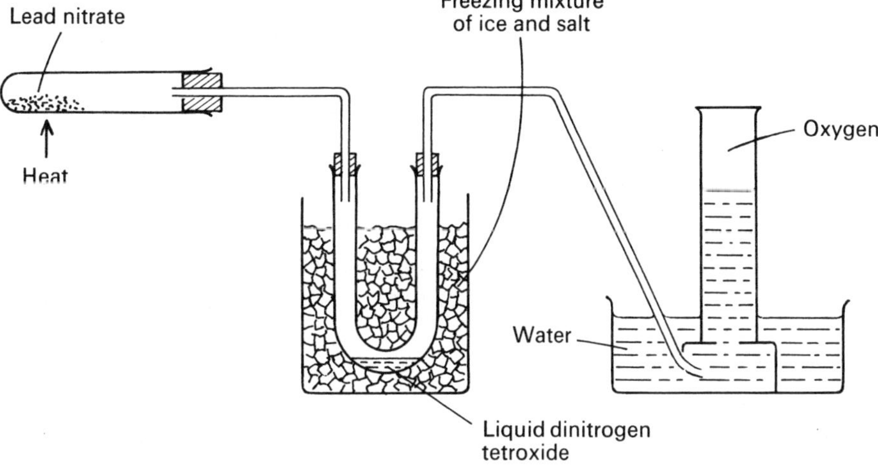

Fig. 29.1 Action of heat on lead nitrate crystals

30 Agricultural chemistry

30.1 Elements necessary for plant growth

The use of chemicals for agricultural purposes is one of the major applications of Chemistry.

In order to grow healthy plants various elements are required in the soil. Nitrogen, phosphorus, potassium, magnesium, calcium and sulphur are required in large amounts. Iron, boron and copper are examples of elements that are required only in very small amounts. These are called **trace elements**.

If soil is used for growing plants year after year, it is possible that the soil could become deficient in certain elements. The soil can be replenished by adding fertilisers.

30.2 Nitrogen fertilisers

Nitrogen is required in large amounts by plants. It is absorbed from the soil through the roots in the form of soluble nitrates. These nitrates are required to build up proteins in the plant. Unit 26.9 explains the circulation of nitrogen in nature, called the nitrogen cycle.

Nitrogen can be supplied to the soil in the form of manure or in the form of dried blood. These natural forms of nitrogen often improve the quality of the soil. There are natural deposits of sodium nitrate in the desert areas of Chile.

Because of the insufficient supply of natural sources of nitrogen, it is necessary to supplement these with artificial fertilisers. These include **calcium nitrate, ammonium sulphate, ammonium nitrate** and **urea**. If a compound is very soluble in water (*e.g.* ammonium nitrate) it is readily washed out of the soil by rain, but before it is washed out its effects are rapid. Urea is insoluble in water but it reacts very slowly with water to produce ammonium compounds. It is therefore suitable as a long-term fertiliser.

When considering which nitrogen fertiliser is most suitable in a particular situation, the following should be considered:

(*i*) percentage of nitrogen in the fertiliser (see 12.9);
(*ii*) cost of the fertiliser;
(*iii*) solubility in water.

For example, Table 30.1 compares information concerning three fertilisers.

Table 30.1 Comparison of three fertilisers

Compound	*Formula*	*Mass of 1 mole*	*Price per kg*	*Solubility in water*
Ammonium nitrate	NH_4NO_3	80 g	50 p	Readily soluble
Calcium cyanamide	$CaCN_2$	80 g	60 p	Insoluble but reacting very slowly
Urea	$CO(NH_2)_2$	60 g	55 p	Insoluble but reacting very slowly

1. Calculate the percentage of nitrogen in ammonium nitrate (see 12.9).
2. Which fertiliser is most suitable for applying in spring so that it will continue to act throughout the summer and autumn?

Obviously ammonium nitrate is not suitable because it is too soluble. The choice between the other two depends on price and the percentage of nitrogen. On this basis, urea (which contains a greater percentage of nitrogen) would be chosen.

30.3 Phosphorus fertilisers

Plants need phosphorus from the soil in order to produce a good root system. This is necessary before a healthy plant can develop.

Phosphorus can be supplied to the soil by **slag** (see 15.7) or **bone meal**. Natural deposits of calcium phosphate $Ca_3(PO_4)_2$ are not very suitable because it is insoluble. However, if calcium phosphate is treated with concentrated sulphuric acid, **calcium superphosphate** is formed.

$$Ca_3(PO_4)_2(s) + 2H_2SO_4(l) \rightarrow Ca(H_2PO_4)_2(s) + 2CaSO_4(s)$$
calcium phosphate + sulphuric acid → calcium superphosphate + calcium sulphate

Calcium superphosphate contains soluble phosphates.

Ammonium phosphate $(NH_4)_3PO_4$ is a suitable soluble phosphorus fertiliser and it also contains nitrogen.

30.4 Potassium fertilisers

Plants need potassium for the production of flowers and seeds. Potassium can be added to the soil in the form of **wood ash** or by the addition of **potassium sulphate**.

30.5 Use of fertilisers

Ready mixed fertilisers are sometimes called **NPK fertilisers** because they supply nitrogen, phosphorus and potassium.

When fertilisers are used, they are sprinkled over the soil or sometimes injected into the soil (*e.g.* liquid ammonia). It is necessary to monitor the pH of the soil. Soils tend to become more acidic as soluble alkalis are washed out of the soil. The use of ammonium sulphate can make the soil more acidic. **Lime** (calcium hydroxide) can be used to neutralise the soil if it is acidic. However, lime and ammonium sulphate should not be used together as ammonia gas is released.

$$(NH_4)_2SO_4(s) + Ca(OH)_2(s) \rightarrow CaSO_4(s) + 2NH_3(g) + 2H_2O(l)$$
ammonium sulphate + calcium hydroxide → calcium sulphate + ammonia + water

31 Phosphorus

31.1 Occurrence

Phosphorus occurs in the earth as calcium phosphate $Ca_3(PO_4)_2$.

31.2 Industrial production of phosphorus

Phosphorus is produced by heating together calcium phosphate, sand (SiO_2) and coke in a furnace. The heating is provided by an electric arc between two carbon electrodes. (This is not an electrolysis process but a reduction process). The furnace is shown in Fig. 31.1.

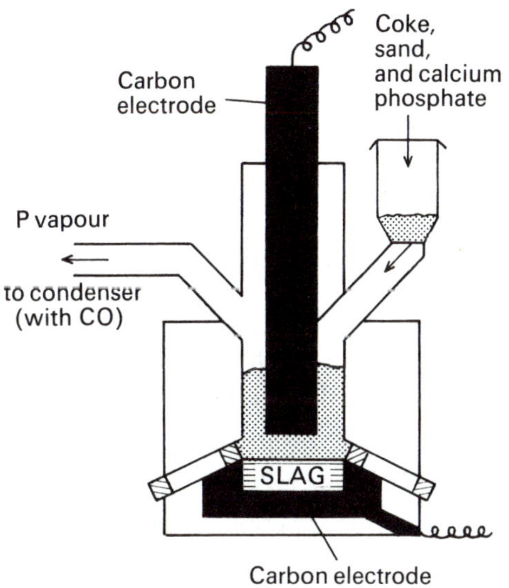

Fig. 31.1 Extraction of phosphorus

The reactions taking place are:

$$2Ca_3(PO_4)_2(s) + 6SiO_2(s) \rightarrow 6CaSiO_3(l) + P_4O_{10}(g)$$

calcium phosphate + silicon dioxide → calcium silicate + phosphorus(V) oxide

$$P_4O_{10}(g) + 10C(s) \rightarrow P_4(g) + 10CO(g)$$

phosphorus(V) oxide + carbon → phosphorus + carbon monoxide

In the second reaction, phosphorus(V) oxide (phosphorus pentoxide) is reduced to phosphorus by carbon. The phosphorus and carbon monoxide gases escape together (preventing oxidation of the phosphorus). The phosphorus condenses under water to form yellow phosphorus.

31.3 Allotropes of phosphorus

There are two common allotropes of phosphorus:—white (yellow) phosphorus and red phosphorus. White phosphorus has a molecular structure of P_4 tetrahedra and red phosphorus has a giant structure of phosphorus atoms. The difference in structure explains the difference in the physical and chemical properties of the two allotropes.

White phosphorus is more readily formed at low temperatures. It is stored under cold water. In excess air, it ignites spontaneously to produce phosphorus(V) oxide (phosphorus pentoxide).

$$P_4(s) + 5O_2(g) \rightarrow P_4O_{10}(s)$$

phosphorus + oxygen → phosphorus(V) oxide

White phosphorus is converted to red phosphorus by heating at about 250°C in an inert atmosphere with a small amount of iodine present as a catalyst.

Red phosphorus reacts with air in a similar way but it has to be heated to a higher temperature. It is not stored under water. It is generally less reactive than white phosphorus. Red phosphorus is converted to white phosphorus by heating in an inert atmosphere to 431°C when the phosphorus sublimes and the solid produced is yellow phosphorus.

31.4 Uses of phosphorus

Phosphorus as an element has few uses. However, phosphorus has many uses in the form of compounds. Phosphorus in the form of phosphates is essential for plant growth (see 30.1). Phosphates are used as water softeners (see 11.4). Phosphorus sulphide, P_4S_3, is used to make matches.

---

32 Hydrogen chloride and chlorine

This unit is written in two parts:

Part A 32.1–32.7
Part B 32.8–32.11

32.1 Hydrogen chloride

When concentrated sulphuric acid is added to sodium chloride (salt) crystals, steamy fumes of hydrogen chloride are formed.

$$NaCl(s) + H_2SO_4(l) \rightarrow NaHSO_4(s) + HCl(g)$$

sodium chloride + sulphuric acid → sodium hydrogensulphate + hydrogen chloride

(Frequently candidates in examinations give the following equation:

$$2NaCl(s) + H_2SO_4(l) \rightarrow Na_2SO_4(s) + 2HCl(g)$$

This equation is unlikely under laboratory conditions.)

Fig. 32.1 shows the apparatus for the preparation of hydrogen chloride. The gas is collected by downward delivery because it is denser than air and soluble in water.

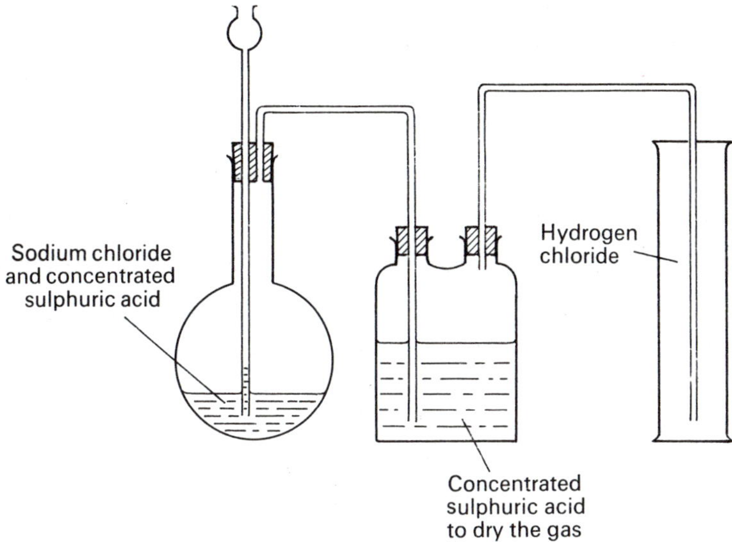

Fig. 32.1 Preparation of hydrogen chloride

Hydrogen chloride is sometimes called **'salt gas'**. It turns blue litmus paper red and when hydrogen chloride fumes are mixed with ammonia gas, a dense white smoke of ammonium chloride is produced.

$$NH_3(g) + HCl(g) \rightarrow NH_4Cl(s)$$
ammonia + hydrogen chloride → ammonium chloride

32.2 To show that salt gas contains hydrogen

If dry hydrogen chloride is passed over heated iron filings, iron(II) chloride is produced. The other product is hydrogen gas which must have come from the salt gas.

$$Fe(s) + 2HCl(g) \rightarrow FeCl_2(s) + H_2(g)$$
iron + hydrogen chloride → iron(II) chloride + hydrogen
(green)

N.B. Iron and chlorine react under similar conditions to form only iron(III) chloride.

32.3 To show that salt gas contains chlorine

If dry hydrogen chloride gas is passed over heated lead(IV) oxide (lead dioxide), chlorine gas is produced. The chlorine must have come from the salt gas.

$$PbO_2(s) + 4HCl(g) \rightarrow PbCl_2(s) + 2H_2O(g) + Cl_2(g)$$
lead(IV) oxide + hydrogen chloride → lead(II) chloride + water + chlorine

32.4 Formula of hydrogen chloride

Electrolysis of a concentrated solution of hydrogen chloride produces hydrogen (at the negative electrode) and chlorine (at the positive electrode). The volumes of hydrogen and chlorine, measured under the same conditions, are the same.

A mixture of hydrogen and chlorine react explosively to form hydrogen chloride.

$$H_2(g) + Cl_2(g) \rightarrow 2HCl(g)$$
hydrogen + chlorine → hydrogen chloride

It is found by experiment that equal volumes of hydrogen and chlorine react together (volumes measured under the same conditions).

32.5 Preparation of hydrochloric acid

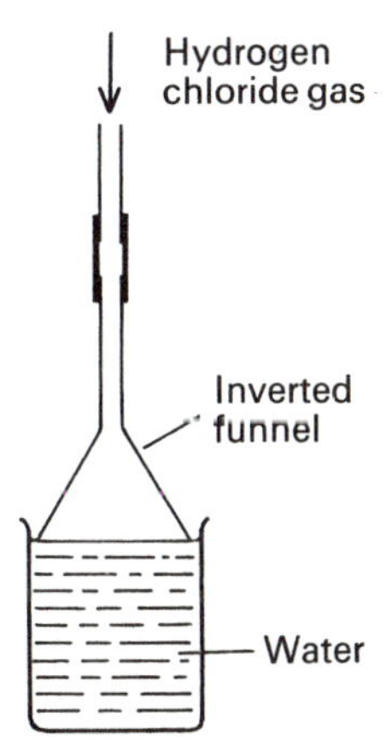

Fig. 32.2 Dissolving hydrogen chloride in water

Hydrochloric acid is prepared when hydrogen chloride is dissolved in water. The apparatus in Fig. 32.2 should be used because hydrogen chloride is very soluble in water. During the process the covalent bond in the hydrogen chloride molecule is broken and ions are formed. The temperature of the solution rises as the ionisation takes place.

$$HCl(g) \rightarrow H^+(aq) + Cl^-(aq)$$

32.6 Properties of concentrated hydrochloric acid

Concentrated hydrochloric acid can be oxidised by a variety of oxidising agents to chlorine.

E.g. $$MnO_2(s) + 4HCl(aq) \rightarrow MnCl_2(aq) + 2H_2O(l) + Cl_2(g)$$
manganese(IV) oxide + hydrochloric acid → manganese(II) chloride + water + chlorine

The mixture has to be heated.

32.7 Properties of dilute hydrochloric acid:

(i) with indicators
A solution of dilute hydrochloric acid turns blue litmus red.

(ii) with metals
Dilute hydrochloric acid reacts with magnesium, zinc and iron to produce hydrogen.

$$Mg(s) + 2HCl(aq) \rightarrow MgCl_2(aq) + H_2(g)$$
magnesium + hydrochloric acid → magnesium chloride + hydrogen

$$Zn(s) + 2HCl(aq) \rightarrow ZnCl_2(aq) + H_2(g)$$
zinc + hydrochloric acid → zinc chloride + hydrogen

$$Fe(s) + 2HCl(aq) \rightarrow FeCl_2(aq) + H_2(g)$$
iron + hydrochloric acid → iron(II) chloride + hydrogen

(Hydrochloric acid does not react, either dilute or concentrated, with copper.)

(iii) with carbonates
Dilute hydrochloric acid reacts with a carbonate, *e.g.* sodium carbonate, to produce carbon dioxide.

$$Na_2CO_3(s) + 2HCl(aq) \rightarrow 2NaCl(aq) + H_2O(l) + CO_2(g)$$
sodium carbonate + hydrochloric acid → sodium chloride + water + carbon dioxide

(iv) with metal oxides, e.g. copper(II) oxide

Dilute hydrochloric acid reacts with copper(II) oxide to form a green solution of copper(II) chloride.

$$CuO(s) + 2HCl(aq) \rightarrow CuCl_2(aq) + H_2O(l)$$
copper(II) oxide + hydrochloric acid → copper(II) chloride + water

32.8 Chlorine

Chlorine is a member of the halogen family (see 18.3). It is a greenish-yellow gas produced by the action of manganese(IV) oxide on warmed, concentrated hydrochloric acid. Suitable apparatus is shown in Fig. 32.3.

The impure chlorine is passed through water (to remove hydrogen chloride fumes) dried by passing through concentrated sulphuric acid and collected by downward delivery.

$$MnO_2(s) + 4HCl(aq) \rightarrow MnCl_2(aq) + 2H_2O(l) + Cl_2(g)$$
manganese(IV) oxide + hydrochloric acid → manganese(II) chloride + water + chlorine

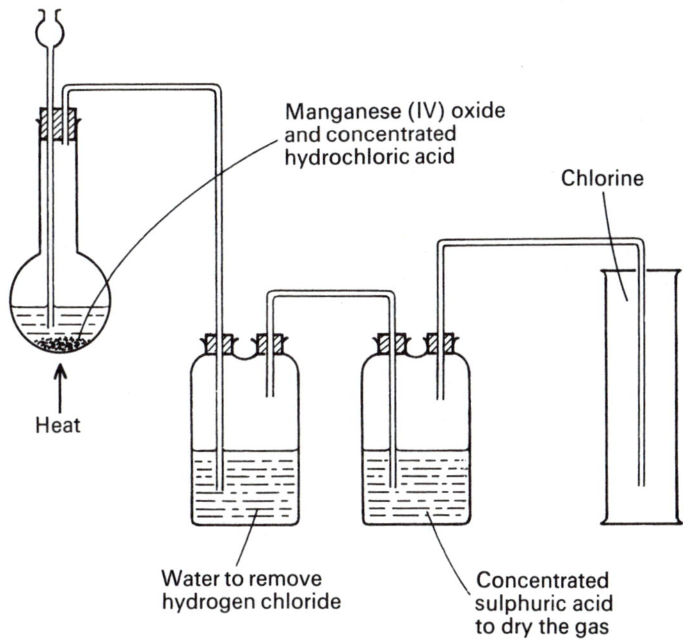

Fig. 32.3 Preparation of chlorine

32.9 Test for chlorine

When a piece of damp blue litmus paper is put into chlorine, the litmus paper turns red and then is bleached.

Chlorine dissolves in water (forming chlorine water) which is slightly acidic because of the reaction of chlorine with water.

$$Cl_2(g) + H_2O(l) \rightleftharpoons HCl(aq) + HClO(aq)$$
chlorine + water ⇌ hydrochloric acid + chloric(I) acid (hypochlorous acid)

The chloric(I) acid is a strong bleaching agent. It decomposes on exposure to sunlight to form oxygen. Chlorine bleaches only in the presence of water. The bleaching occurs by oxidation of the dye with the decomposition of chloric(I) acid. (N.B. Sulphur dioxide bleaches by reduction, see 24.5.)

32.10 Reactions of chlorine

Chlorine is a very reactive element. It reacts directly with all elements except carbon, nitrogen, oxygen and the inert gases.

(i) with metals

When chlorine is passed over a heated metal powder, the anhydrous metal chloride is formed (see 18.3).

E.g.

$$2Fe(s) + 3Cl_2(g) \rightarrow 2FeCl_3(s)$$
iron + chlorine → iron(III) chloride

(ii) with compounds containing hydrogen

Chlorine has a great affinity for hydrogen and oxidises many compounds by removing hydrogen (see 16.1).

E.g. 1. Hydrogen sulphide

If hydrogen sulphide and chlorine are mixed in the presence of water, the hydrogen sulphide is oxidised to sulphur.

$$H_2S(g) + Cl_2(g) \rightarrow 2HCl(g) + S(s)$$
hydrogen sulphide + chlorine → hydrogen chloride + sulphur

E.g. 2. Turpentine
Turpentine is a hydrocarbon ($C_{10}H_{16}$). If warm turpentine is put into chlorine, the turpentine is oxidised to carbon.

$$C_{10}H_{16}(l) + 8Cl_2(g) \rightarrow 10C(s) + 16HCl(g)$$
turpentine + chlorine → carbon + hydrogen chloride

E.g. 3. Wax taper
The wax used to make a wax taper is a mixture of hydrocarbons and can be represented as C_xH_y. A lighted wax taper burns with a red smoky flame when put into a gas jar of chlorine. The products are carbon and hydrogen chloride.

(iii) with alkalis, e.g. sodium hydroxide
The reaction of chlorine with an alkali, *e.g.* sodium hydroxide, depends upon the conditions. When chlorine is passed into a cold dilute solution of sodium hydroxide, a mixture of sodium chloride and sodium chlorate(I) (sodium hypochlorite) is formed.

$$2NaOH(aq) + Cl_2(g) \rightarrow NaCl(aq) + NaClO(aq) + H_2O(l)$$
sodium hydroxide + chlorine → sodium chloride + sodium chlorate(I) + water

When chlorine is passed into hot, concentrated sodium hydroxide solution, a mixture of sodium chloride and sodium chlorate(V) (sodium chlorate) is formed.

$$6NaOH(aq) + 3Cl_2(g) \rightarrow 5NaCl(aq) + NaClO_3(aq) + 3H_2O(l)$$
sodium hydroxide + chlorine → sodium chloride + sodium chlorate(V) + water

Bleaching powder is produced when chlorine is passed over cold, solid calcium hydroxide (slaked lime).

(iv) displacement reactions (see 18.3)

32.11 Manufacture and uses of chlorine

Chlorine is manufactured during the electrolysis of chlorides, *e.g.* sodium chloride (see 14.3 and 42).

It is widely used for bleaching certain fabrics and woodpulp. It is also used for killing germs, for making hydrochloric acid from hydrogen and chlorine, and in a wide range of organic compounds including pesticides (DDT), synthetic fibres and plastics (PVC).

---

33 Carbon, carbon monoxide and organic chemistry

This unit is written in three parts:

Part A	33.1–33.3	Carbon
Part B	33.4–33.6	Carbon monoxide
Part C	33.7–33.28	Organic chemistry

Check in Section I to find which parts are required for your syllabus.

33.1 Occurrence of carbon

The element carbon occurs naturally in the form of diamond and graphite and in an impure form as coal. It also occurs combined with other elements in petroleum, metal carbonates and in the atmosphere in the form of carbon dioxide.

33.2 Allotropy of carbon

Two crystalline allotropes exist—diamond and graphite. Other forms of carbon are known collectively as amorphous carbon.

(i) Diamond

In the diamond structure each carbon atom is strongly bound (covalent bonding) to four other carbon atoms tetrahedrally. A large giant structure (3 dimensional) is built up. All bonds between carbon atoms are the same length (0.154 nm). It is the strength and uniformity of the bonding which make diamond very hard, non-volatile and resistant to chemical attack. Fig. 33.1 shows the arrangement of particles in diamond.

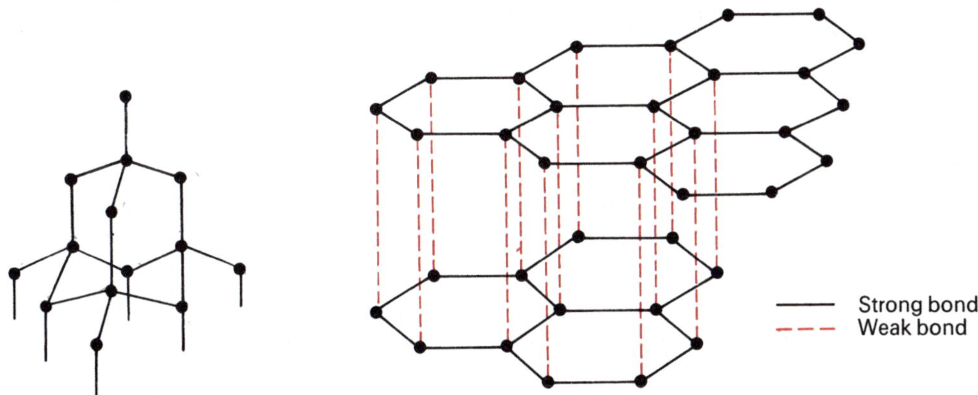

Fig. 33.1 Structure of diamond

Fig. 33.2 Structure of graphite

(ii) Graphite

Graphite has a layer structure. In each layer the carbon atoms are bound covalently. The bonds within the layers are very strong. The bonds between the layers, however, are very weak which enables layers to slide over one another. This makes the graphite soft and flaky. Fig. 33.2 shows the arrangement of particles in graphite.

Table 33.1 compares the properties of diamond and graphite.

Table 33.1 Comparing the properties of diamond and graphite

Property	*Diamond*	*Graphite*
Appearance	Transparent, colourless crystals	Black, opaque, shiny solid
Density (g/cm^3)	3.5	2.2
Hardness	Very hard	Very soft
Electrical conductivity	Non-conductor	Good electrical conductor
Burning in oxygen	Burns only with difficulty when heated to high temperature. Carbon dioxide produced. No residue	Burns readily to produce carbon dioxide. No residue

(iii) Amorphous carbon

Amorphous carbon is made up of minute particles of graphite and exists in several forms including wood charcoal, animal charcoal and lampblack. Coke, soot and gas carbon are other forms of impure amorphous carbon.

33.3 Properties of carbon

(i) Combustion

Carbon burns in excess oxygen to form carbon dioxide, the temperature at which combustion starts depending on the form of carbon.

$$C(s) + O_2(g) \rightarrow CO_2(g)$$

$$\text{carbon} + \text{oxygen} \rightarrow \text{carbon dioxide}$$

(ii) As a reducing agent (see 16.2)
Carbon will reduce certain metal oxides to the metal.

E.g.
$$PbO(s) + C(s) \rightarrow Pb(s) + CO(g)$$
lead(II) oxide + carbon → lead + carbon monoxide

33.4 LABORATORY PREPARATIONS OF CARBON MONOXIDE

(i) from sodium methanoate (sodium formate)
Carbon monoxide is prepared from sodium methanoate and concentrated sulphuric acid using the apparatus in Fig. 33.3.

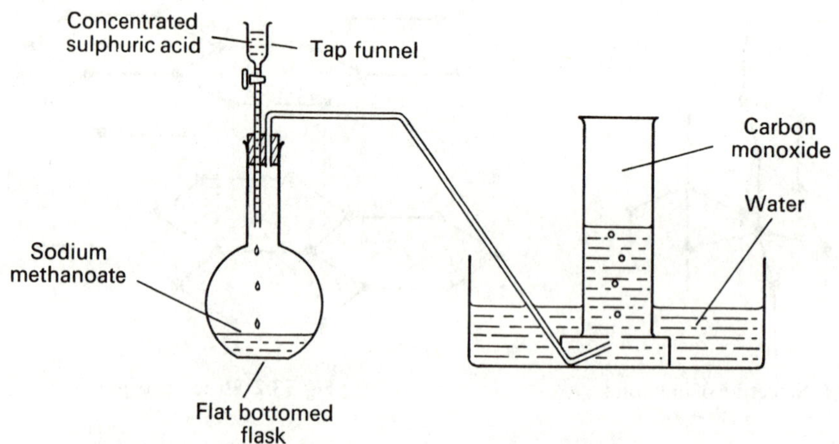

Fig. 33.3 Preparation of carbon monoxide from sodium methanoate

Two reactions are involved:
Sodium methanoate is converted to methanoic acid (formic acid) by the sulphuric acid.

$$HCOONa(s) + H_2SO_4(l) \rightarrow HCOOH(l) + NaHSO_4(s)$$
sodium methanoate + sulphuric acid → methanoic acid + sodium hydrogensulphate

Concentrated sulphuric acid removes water from the methanoic acid.

$$HCOOH(l) \rightarrow H_2O(l) + CO(g)$$
methanoic acid → water + carbon monoxide

Overall these reactions can be represented:

$$HCOONa(s) + H_2SO_4(l) \rightarrow NaHSO_4(s) + CO(g) + H_2O(l)$$
sodium methanoate + sulphuric acid → sodium hydrogensulphate + carbon monoxide + water

(ii) from ethanedioic acid (oxalic acid)
Carbon monoxide can also be prepared by the dehydration of ethanedioic acid with concentrated sulphuric acid.

$$H_2C_2O_4(s) \rightarrow CO(g) + CO_2(g) + H_2O(l)$$
ethanedioic acid → carbon monoxide + carbon dioxide + water

The mixture of carbon monoxide and carbon dioxide is passed through potassium hydroxide solution to absorb the carbon dioxide. The carbon monoxide can be collected over water or, if required dry, can be dried by passing through concentrated sulphuric acid and collected over mercury.

33.5 PROPERTIES OF CARBON MONOXIDE

(*i*) Carbon monoxide is a colourless, odourless gas which is only very slightly soluble in water.

(*ii*) It is extremely poisonous since it combines irreversibly with the haemoglobin in the blood to form the compound carboxyhaemoglobin. This prevents the haemoglobin carrying oxygen. An atmosphere containing as little as 0.5% carbon monoxide may cause death if breathed for some time. Since exhaust fumes contain carbon monoxide it is dangerous to run a car engine in a garage.

(*iii*) Carbon monoxide is a good reducing agent (see 16.2). It will reduce many metal oxides to the metal on heating.

E.g.

$$PbO(s) + CO(g) \rightarrow Pb(s) + CO_2(g)$$
lead(II) oxide + carbon monoxide → lead + carbon dioxide

$$Fe_2O_3(s) + 3CO(g) \rightarrow 2Fe(s) + 3CO_2(g)$$
iron(III) oxide + carbon monoxide → iron + carbon dioxide

$$CuO(s) + CO(g) \rightarrow Cu(s) + CO_2(g)$$
copper(II) oxide + carbon monoxide → copper + carbon dioxide

(*iv*) Carbon monoxide burns with a bright blue flame producing carbon dioxide. (The carbon dioxide may be detected by testing with limewater.)

$$2CO(g) + O_2(g) \rightarrow 2CO_2(g)$$
carbon monoxide + oxygen → carbon dioxide

This is used as the test for carbon monoxide.

33.6 Uses of carbon monoxide

(*i*) As a fuel (see 36.4)
(*ii*) In the manufacture of methanol
(*iii*) In the manufacture of synthetic petrol
(*iv*) In the reduction of ores and the refining of nickel

33.7 Organic chemistry

Organic chemistry is the chemistry of the compounds of carbon with the exception of the oxides (carbon monoxide and carbon dioxide), metal carbonates, the disulphide (carbon disulphide) and the tetrachloride (tetrachloromethane) which are conventionally regarded as being inorganic and are studied in that branch of chemistry.

Carbon is able to form a very large number of compounds because it forms stable covalent bonds with other atoms. In fact, carbon shows exceptional behaviour in that it is able to form chains or rings of great size and complexity.

33.8 Organic formulae

There are two main ways of representing the formula of an organic compound.

(i) Molecular formula
The molecular formula gives information about which elements are present in the organic compound. It also indicates the number of atoms of each element which are present in one molecule of the compound.

E.g. C_2H_5ON is the molecular formula of ethanamide (acetamide)—an organic compound.

(ii) Structural formula
In organic chemistry the molecular formula is often not sufficient to identify a particular organic compound. This is because two different organic compounds can exist which have the same molecular formula but different arrangements of atoms. This is called **isomerism.** To overcome this problem a formula, known as the structural formula, is used. It is a two-dimensional representation of the compound which gives information on the arrangement of atoms within the molecule. Fig. 33.4 shows the structural formula of ethanamide.

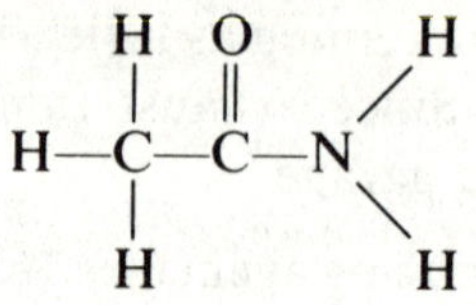

Fig. 33.4 Structural formula of ethanamide (acetamide)

It should be noted that:

(*i*) each carbon atom forms 4 bonds with other atoms;
(*ii*) nitrogen atoms form 3 bonds with other atoms;
(*iii*) oxygen atoms form 2 bonds with other atoms;
(*iv*) hydrogen atoms form one bond with another atom.

33.9 ALKANES

The alkanes are a series of hydrocarbons (compounds of carbon and hydrogen) with the general formula C_nH_{2n+2} where n = 1,2,3,..., for successive members of the series.

The first member of the series (n = 1) is methane CH_4 and the second member (n = 2) is ethane C_2H_6.

Table 33.2 summarises some information about the simplest alkanes.

Table 33.2 The simplest alkanes

Alkane	*Molecular formula*	*Structural formula*	*Melting point °C*	*Boiling point °C*	*State at room temperature and pressure*
Methane	CH_4	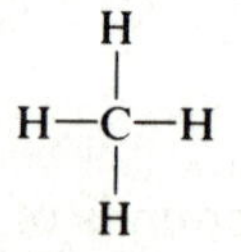	−183	−162	gas
Ethane	C_2H_6	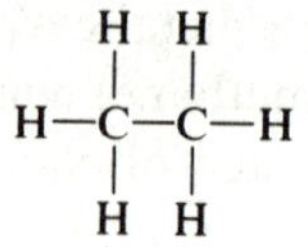	−183	−89	gas
Propane	C_3H_8	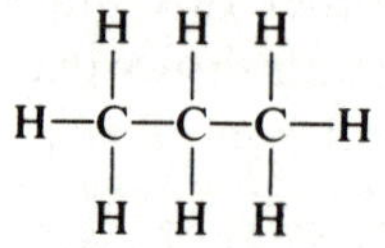	−188	−42	gas
Butane	C_4H_{10}	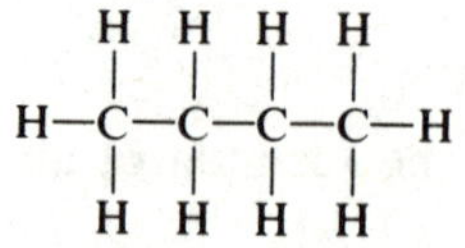	−135	−0.5	gas
Pentane	C_5H_{12}	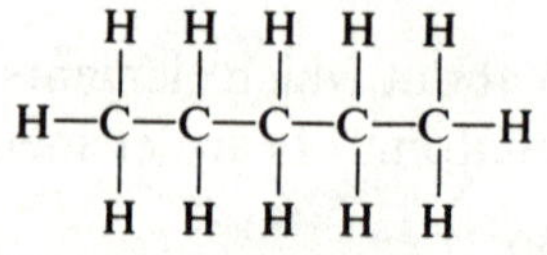	−130	36	liquid

Alkanes contain only single bonds and are said to be **saturated**. Organic compounds containing double or triple bonds are said to be **unsaturated**.

A series of compounds which are related to each other (*e.g.* the alkanes) is called a **homologous series**. Each member is called a **homologue**. In each homologous series, each member has the same general formula but differs from the next in the series by a unit of CH_2.

The physical properties of the members show a gradual change with increasing relative molecular mass. Thus, for example, in the alkanes the melting points and boiling points rise with increasing relative molecular mass. In a similar way, the densities increase and the solubility in water decreases with rising relative molecular mass.

33.10 LABORATORY PREPARATION OF METHANE

Methane can be prepared by heating a mixture of anhydrous sodium ethanoate (sodium acetate) and soda-lime (a non-deliquescent form of sodium hydroxide). The apparatus is shown in Fig. 33.5.

$$CH_3COONa(s) + NaOH(s) \rightarrow Na_2CO_3(s) + CH_4(g)$$
sodium ethanoate + sodium hydroxide → sodium carbonate + methane

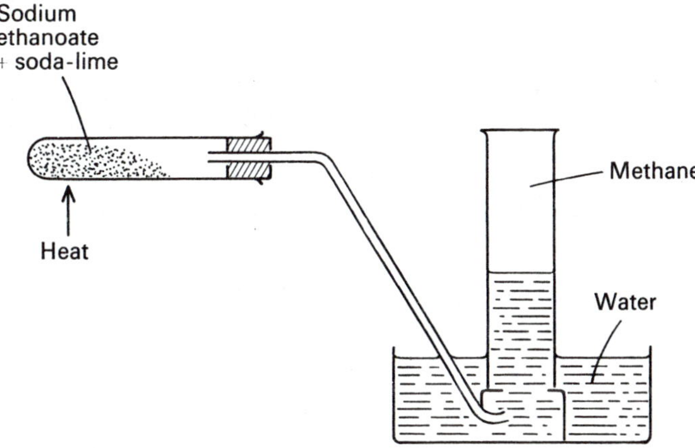

Fig. 33.5 Preparation of methane from sodium ethanoate

33.11 CHEMICAL REACTIONS OF METHANE

(i) Methane burns in air or oxygen
It forms carbon dioxide in excess air and carbon monoxide in a limited supply of air.

$$CH_4(g) + 2O_2(g) \rightarrow CO_2(g) + 2H_2O(g)$$
methane + oxygen → carbon dioxide + water

$$2CH_4(g) + 3O_2(g) \rightarrow 2CO(g) + 4H_2O(g)$$
methane + oxygen → carbon monoxide + water

(ii) Reaction with chlorine
In diffused light, methane reacts giving a series of products by the successive replacement of hydrogen atoms. Each of the reactions is a **substitution reaction.**

$$CH_4(g) + Cl_2(g) \rightarrow HCl(g) + CH_3Cl(g)$$
methane + chlorine → hydrogen chloride + chloromethane

$$CH_3Cl(g) + Cl_2(g) \rightarrow HCl(g) + CH_2Cl_2(g)$$
chloromethane + chlorine → hydrogen chloride + dichloromethane

$$CH_2Cl_2(g) + Cl_2(g) \rightarrow HCl(g) + CHCl_3(g)$$
dichloromethane + chlorine → hydrogen chloride + trichloromethane

$$CHCl_3(g) + Cl_2(g) \rightarrow HCl(g) + CCl_4(g)$$
trichloromethane + chlorine → hydrogen chloride + tetrachloromethane

33.12 ISOMERISM

There is only one possible structure for each of the first three alkanes but the four carbon atoms and the ten hydrogen atoms in a molecule of butane, C_4H_{10}, can be linked in two different ways as shown in Fig. 33.6.

```
     H  H  H  H              H  H  H
     |  |  |  |              |  |  |
  H—C—C—C—C—H           H—C—C—C—H
     |  |  |  |              |  |  |
     H  H  H  H              H  |  H
                                 H—C—H
        Butane                      |
    (or n-Butane)                   H

                           2-methylpropane
                           (or iso-butane)
```

Fig. 33.6 Isomers of butane

This is an example of isomerism, *i.e.* the existence of two or more compounds (called isomers) having the same molecular formula but different structural formulae.

Isomerism becomes more common in the higher alkanes, *e.g.* there are 75 isomers of decane $C_{10}H_{22}$. Isomerism also occurs within other homologous series and also with compounds in different homologous series:

E.g. C_2H_5OH (an alcohol) and CH_3OCH_3 (an ether)

33.13 Alkenes

The alkenes are a series of hydrocarbons with the general formula C_nH_{2n}, where n = 2, 3,.... Table 33.3 summarises some information about the simplest alkenes.

Table 33.3 The simplest alkenes

Alkene	*Molecular formula*	*Structural formula*	*Melting point °C*	*Boiling point °C*	*State at room temperature and pressure*
Ethene	C_2H_4	H H \\ / C=C / \\ H H	−169	−102	gas
Propene	C_3H_6	H | H—C H | \\ / H C=C / \\ H H	−185	−48	gas
Butene	C_4H_8	H H | | H—C—C H | | \\ / H H C=C / \\ H H	−185	−6	gas
Pentene	C_5H_{10}	H H H | | | H—C—C—C H | | | \\ / H H H C=C / \\ H H	−138	30	liquid

It should be noted that all alkenes contain a double bond between two carbon atoms. They are, therefore, **unsaturated**. Alkenes with four or more carbon atoms can exist as different isomers.

33.14 Preparation of Ethene

Ethene is prepared by the dehydration of ethanol by hot, concentrated sulphuric acid using the apparatus in Fig. 33.7.

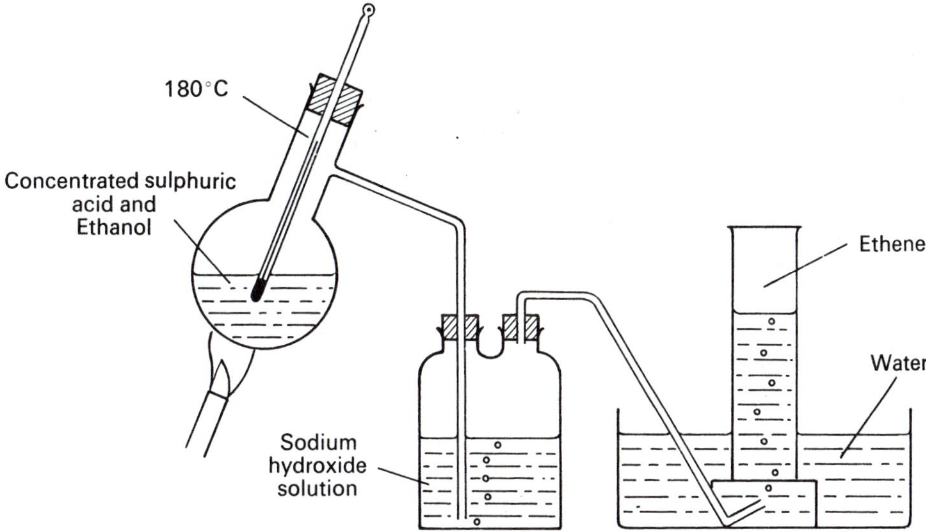

Fig. 33.7 Preparation of ethene

The ethene evolved is passed through sodium hydroxide solution to remove sulphur dioxide which is produced as a by-product. The ethene is then collected over water as it is almost insoluble.

Two reactions are involved:

(*i*) Ethanol reacts with concentrated sulphuric acid to form ethyl hydrogensulphate.

$$C_2H_5OH(l) + H_2SO_4(l) \rightarrow C_2H_5HSO_4(l) + H_2O(l)$$
ethanol + sulphuric acid → ethyl hydrogensulphate + water

(*ii*) Ethyl hydrogensulphate decomposes on heating

$$C_2H_5HSO_4(l) \rightarrow C_2H_4(g) + H_2SO_4(l)$$
ethyl hydrogensulphate → ethene + sulphuric acid

33.15 Reactions of Ethene

Other alkenes react in a similar way.

(i) Combustion. Ethene burns in air or oxygen if ignited. In excess air carbon dioxide and water are produced and in a limited supply of air carbon monoxide and water are produced.

(ii) Addition reactions. Addition reactions are common with all unsaturated compounds. In such a reaction two substances combine to produce a single new substance. The reaction between ethene and bromine is most frequently mentioned. Ethene reacts rapidly with bromine vapour to form colourless oily drops of 1,2-dibromoethane (Fig. 33.8).

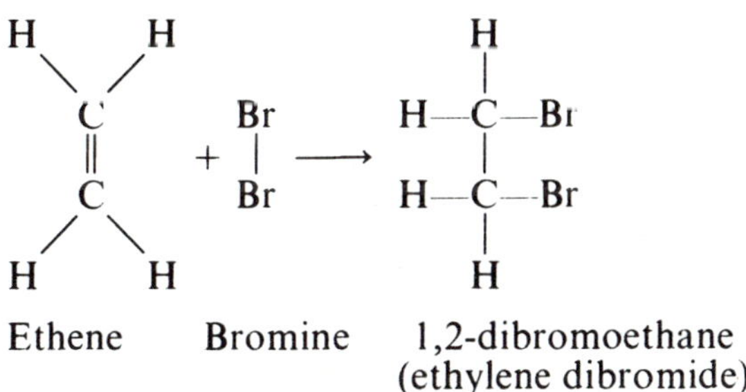

Fig. 33.8 Addition of bromine to ethene

This reaction is used to detect compounds containing double or triple bonds (unsaturated). The compound is shaken with bromine dissolved in a suitable solvent (*e.g.* tetrachloromethane). Unsaturated compounds remove the reddish colour of the bromine. This test does not, of course, distinguish between compounds containing double and triple bonds.

Ethene also reacts with hydrogen at 200°C, in the presence of a finely divided nickel catalyst, to form ethane.

$$C_2H_4(g) + H_2(g) \rightarrow C_2H_6(g)$$
$$\text{ethene} + \text{hydrogen} \rightarrow \text{ethane}$$

This reaction is similar to the reaction in which natural fats and oils are converted to margarine by the addition of hydrogen at about 5 atmospheres pressure and 180°C.

33.16 Alkynes

The alkynes are a series of hydrocarbons with a general formula C_nH_{2n-2}. The simplest member is ethyne (sometimes called acetylene) C_2H_2, which has the structural formula $H{-}C{\equiv}C{-}H$. All alkynes contain a triple bond between two carbon atoms and, like the alkenes, are unsaturated.

33.17 Preparation of ethyne

Ethyne is prepared in the laboratory by the reaction between calcium carbide and water, using the apparatus shown in Fig. 33.9.

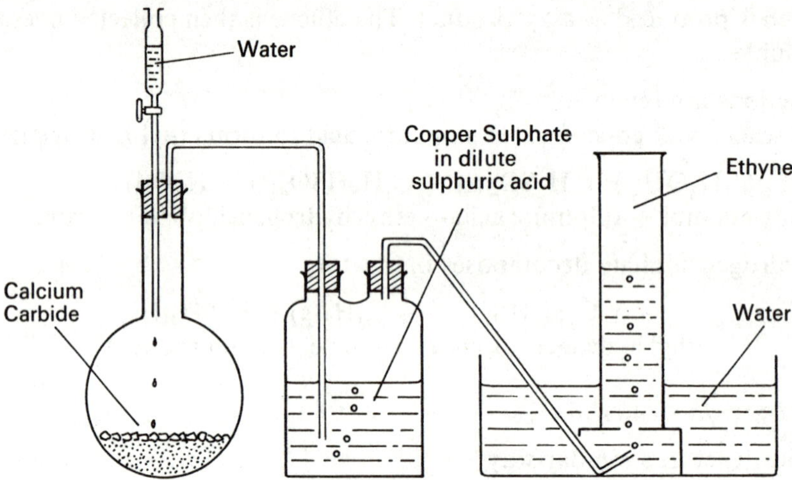

Fig. 33.9 Preparation of ethyne

The ethyne evolved is first passed through a solution of copper(II) sulphate to remove impurities of phosphine (PH_3) and hydrogen sulphide (H_2S), before being collected over water.

$$CaC_2(s) + 2H_2O(l) \rightarrow Ca(OH)_2(s) + C_2H_2(g)$$
$$\text{calcium carbide} + \text{water} \rightarrow \text{calcium hydroxide} + \text{ethyne}$$

33.18 Reactions of ethyne

(i) Combustion. Ethyne burns with a very smoky flame. With a plentiful supply of air or oxygen, carbon dioxide and water are produced

$$2C_2H_2(g) + 5O_2(g) \rightarrow 4CO_2(g) + 2H_2O(g)$$
$$\text{ethyne} + \text{oxygen} \rightarrow \text{carbon dioxide} + \text{water}$$

Issuing from a jet, ethyne burns very exothermically in air producing temperatures up to 3000°C. This is the basis of the oxyacetylene flame which is used in welding and metal cutting.

(ii) Addition reactions. Alkynes undergo addition reactions because they are unsaturated. Ethyne reacts with bromine to form tetrabromoethane.

$$C_2H_2(g) + 2Br_2(g) \rightarrow C_2H_2Br_4$$
$$\text{ethyne} + \text{bromine} \rightarrow \text{tetrabromoethane}$$

33.19 Oil

Crude oil (also called mineral oil or petroleum) is found in many parts of the world. Four areas of production are the Middle East, USA (including Alaska), Venezuela and the North Sea.

It is found trapped in permeable rock between layers of impermeable rock (see Fig. 36.2). It was formed by the decomposition of animal and plant material under pressure. It is extracted by drilling deep holes into the earth. Often the oil will issue from the ground under its own pressure but pumping may be required later.

33.20 Refining of crude oil

Crude oil consists mainly of a complex mixture of hydrocarbons, the greater proportion of which are alkanes. The crude oil can be split up into various fractions by **fractional distillation.**

Fractional distillation of crude oil may be carried out on a small scale in the laboratory using the apparatus in Fig. 33.10.

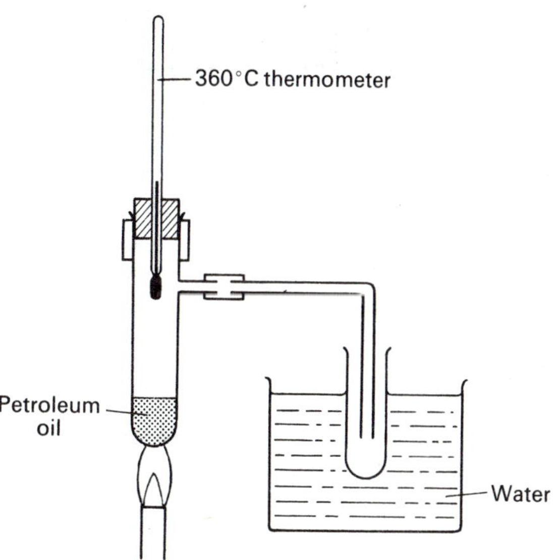

Fig. 33.10 Laboratory distillation of oil

Several fractions may be obtained corresponding to different boiling point ranges (*e.g.* 1st fraction—up to 70°C, 2nd fraction 70°–120°C, 3rd fraction 120°–170°C, 4th fraction 170°–220°C *etc.*). On examining the fractions obtained, certain definite changes in properties of the fractions with increasing boiling point can be seen. In particular, the yellow colour and viscosity (*i.e.* the ease with which the liquid pours) increase with increasing boiling point whilst the inflammability decreases.

Industrially, the fractional distillation of crude oil is carried out on a large scale in an oil refinery. The fractionation is carried out in a fractional distillation column (Fig. 33.11).

The main fractions include: (*i*) petrol (6–12 carbon atoms);
(*ii*) paraffin (11–16 carbon atoms);
(*iii*) lubricating oil (more than 20 carbon atoms).

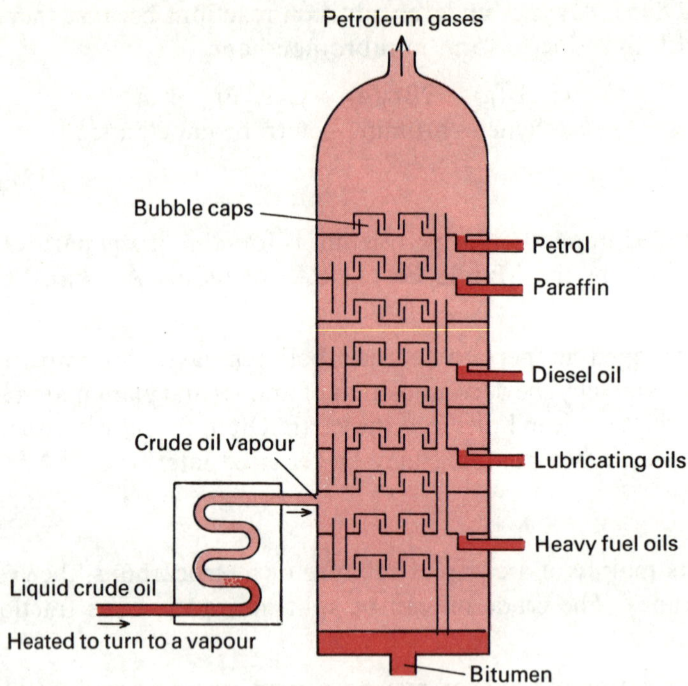

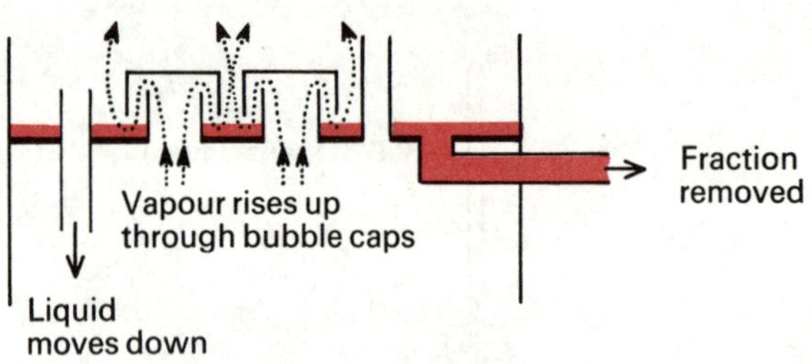

Fig. 33.11 Industrial distillation of oil

33.21 Cracking

Crude oil contains many long chain alkanes. **Cracking** is a process which has been developed by the oil companies to produce larger quantities of the shorter chain hydrocarbons which are easier to market. It involves the breaking of long chain alkanes and can be carried out in two ways:

(*i*) **Thermal cracking**—carried out simply by heating.
(*ii*) **Catalytic cracking**—carried out using a catalyst.

The cracking of a long chain alkane always leads to the formation of at least one unsaturated product (Fig. 33.12).

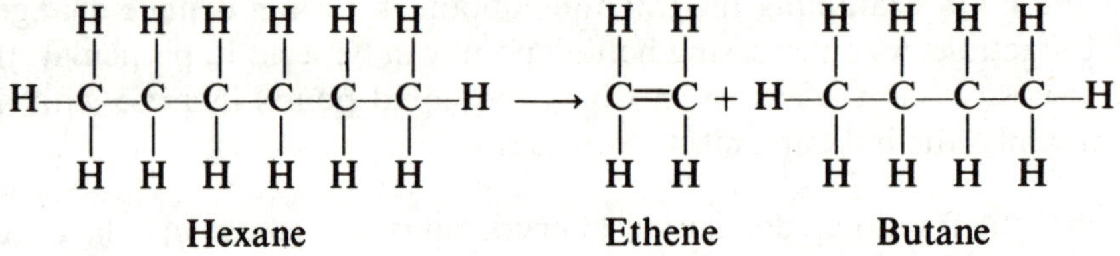

Fig. 33.12 Cracking of hexane

When liquid paraffin vapour is cracked by passing over strongly heated broken china, ethene gas is produced and can be collected over water (Fig. 33.13).

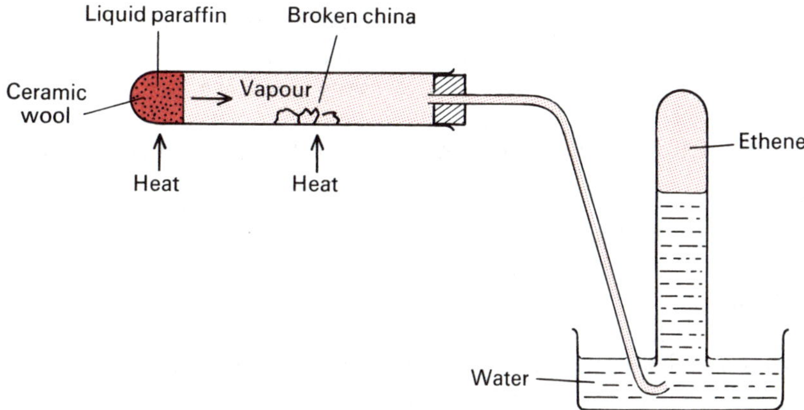

Fig. 33.13 Thermal cracking of liquid paraffin

33.22 ALCOHOLS

Alcohols are compounds of carbon, hydrogen and oxygen. They form an homologous series with the general formula $C_nH_{2n+1}OH$. They may be regarded as derived from an alkane by replacing a hydrogen atom by a hydroxyl (—OH) group. The most important alcohol is ethanol C_2H_5OH.

33.23 PREPARATION OF ETHANOL

Ethanol is prepared by the **fermentation** of glucose using enzymes in yeast. The apparatus (Fig. 33.14) is kept at about 30°C for a couple of days. The fermentation lock allows the escape of carbon dioxide without the entry of oxygen which, in the presence of certain bacteria, would convert the ethanol to a dilute solution of ethanoic acid. This is called **souring.**

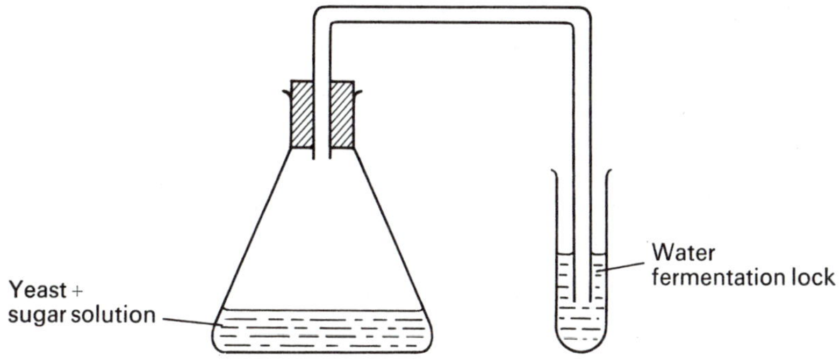

Fig. 33.14 Fermentation

During the fermentation the glucose is converted to ethanol by the enzyme **zymase** in yeast.

$$C_6H_{12}O_6(aq) \rightarrow 2C_2H_5OH(aq) + 2CO_2(g)$$
$$\text{glucose} \rightarrow \text{ethanol} + \text{carbon dioxide}$$

A dilute solution of ethanol is produced and the ethanol can be concentrated by **fractional distillation** (see 2.3).

33.24 PROPERTIES OF ETHANOL

(i) Combustion. When ethanol is ignited in a plentiful supply of air or oxygen the products are carbon dioxide and water.

(ii) Ester formation. Ethanol reacts reversibly with ethanoic acid, in the presence of concentrated sulphuric acid, to form ethyl ethanoate (ethyl acetate). Ethyl ethanoate is an **ester.**

$$CH_3COOH(l) + C_2H_5OH(l) \rightleftharpoons CH_3COOC_2H_5(l) + H_2O(l)$$

ethanoic acid + ethanol ⇌ ethyl ethanoate + water

ACID + ALCOHOL ⇌ ESTER + WATER

Esters are sweet-smelling liquids found naturally in flowers and fruit to which they give scent and flavour.

Many naturally occurring fats and oils consist largely of esters. When an ester is boiled with an alkali solution the ester is split up into the constituent acid and alcohol. This is called **saponification.**

(iii) Dehydration of ethanol with concentrated sulphuric acid produces ethene (see 33.14). This can also be done by passing ethanol vapour over heated aluminium oxide.

33.25 Carboxylic acids

All organic acids contain the carboxyl group — COOH in the molecule. The simplest homologous series of acids has a general formula $C_nH_{2n+1}COOH$. The simplest acids are formic or methanoic acid HCOOH and acetic or ethanoic acid, CH_3COOH.

Ethanoic acid can be prepared by the oxidation of ethanol using acidified potassium dichromate solution. These acids are weak acids.

33.26 Carbohydrates

Carbohydrates are compounds containing carbon, hydrogen and oxygen. The last two elements are present in the same proportion as in water. All carbohydrates have the general molecular formula $C_x(H_2O)_y$.

Carbohydrates may be divided into **monosaccharides** (*e.g.* glucose and fructose, both $C_6H_{12}O_6$), **disaccharides** (*e.g.* sucrose and maltose, both $C_{12}H_{22}O_{11}$) and **polysaccharides** (*e.g.* starch). Monosaccharides and disaccharides are often referred to as sugars.

33.27 Tests for starch and reducing sugars

Starch produces a dark blue coloration with iodine solution. This is used as a test for starch.

Certain sugars will reduce hot **Fehling's** (or **Benedict's**) **solution** to a brick red precipitate of copper(I) oxide. These sugars are called **reducing sugars.** Examples of reducing sugars are glucose, fructose and maltose.

33.28 Hydrolysis of starch

Starch may be broken down into simpler carbohydrates by **hydrolysis** in aqueous solution. The breakdown involves the reaction of the starch with water but unless a catalyst is present the reaction is extremely slow. The reaction can be catalysed in two ways:

(i) Acid catalysed hydrolysis. This is carried out by heating starch solution with dilute acid. The product is **glucose.**

(ii) Enzyme catalysed hydrolysis. The hydrolysis reaction may also be catalysed by the enzyme α-amylase which is present in saliva. This reaction proceeds at room temperature to produce **maltose.**

34 Carbon dioxide and carbonates

34.1 Laboratory preparation of carbon dioxide

Carbon dioxide can be prepared by the reaction between marble chips (calcium carbonate) and dilute hydrochloric acid using the apparatus in Fig. 34.1.

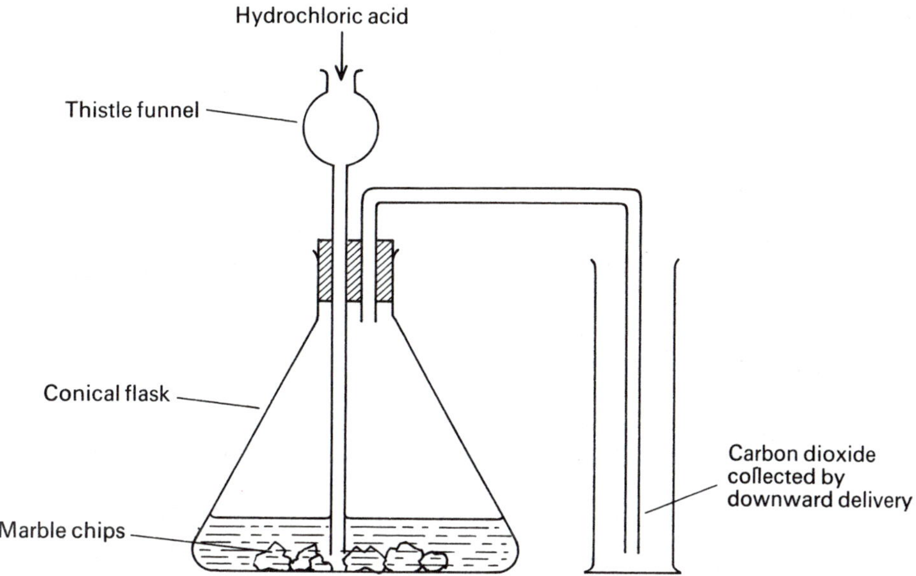

Fig. 34.1 Preparation of carbon dioxide

The carbon dioxide can be collected by downward delivery (because carbon dioxide is much denser than air) or collected over water (although it is quite soluble in water). A solution of calcium chloride remains in the flask.

$$CaCO_3(s) + 2HCl(aq) \rightarrow CaCl_2(aq) + H_2O(l) + CO_2(g)$$
calcium carbonate + hydrochloric acid → calcium chloride + water + carbon dioxide

If a pure, dry sample of carbon dioxide is required it may be passed through a solution of potassium hydrogencarbonate (to remove hydrochloric acid spray which may become suspended in the gas), dried by passing it through concentrated sulphuric acid (see 25.6) and collected by downward delivery.

34.2 Properties of carbon dioxide

Carbon dioxide is a colourless, odourless gas and it is denser than air.

(i) Solubility in water. Carbon dioxide is moderately soluble in water. At room temperature and pressure, water will absorb its own volume of carbon dioxide. A solution of carbon dioxide in water is slightly acidic, having a pH of about 5.5. This is because carbon dioxide reacts with water to produce a weak acid, carbonic acid. Carbon dioxide is the anhydride of carbonic acid (see 9.3).

$$CO_2(g) + H_2O(l) \rightleftharpoons H_2CO_3(aq)$$
carbon dioxide + water ⇌ carbonic acid

(ii) Carbon dioxide as a supporter of combustion. If a lighted splint is plunged into carbon dioxide the splint is extinguished because carbon dioxide does not usually support combustion.

However, carbon dioxide supports the combustion of strongly burning magnesium ribbon because the carbon dioxide is decomposed. Black specks of carbon and white magnesium oxide powder are formed.

$$2Mg(s) + CO_2(g) \rightarrow 2MgO(s) + C(s)$$
magnesium + carbon dioxide → magnesium oxide + carbon

(iii) Carbon dioxide reacts with an aqueous solution of sodium hydroxide to produce a solution of sodium carbonate.

$$2NaOH(aq) + CO_2(g) \rightarrow Na_2CO_3(aq) + H_2O(l)$$
sodium hydroxide + carbon dioxide → sodium carbonate + water

Excess carbon dioxide produces sodium hydrogencarbonate which may be produced as a white precipitate.

$$Na_2CO_3(aq) + H_2O(l) + CO_2(g) \rightarrow 2NaHCO_3(s)$$
sodium carbonate + water + carbon dioxide → sodium hydrogencarbonate

34.3 Test for carbon dioxide

If carbon dioxide is passed into limewater (an aqueous solution of calcium hydroxide) the limewater turns milky due to the formation of a white precipitate of calcium carbonate.

$$Ca(OH)_2(aq) + CO_2(g) \rightarrow CaCO_3(s) + H_2O(l)$$
calcium hydroxide + carbon dioxide → calcium carbonate + water

If excess carbon dioxide is passed into limewater, the limewater will become clear again. This is due to the formation of calcium hydrogen carbonate which is soluble in water.

$$CaCO_3(s) + H_2O(l) + CO_2(g) \rightleftharpoons Ca(HCO_3)_2(aq)$$
calcium carbonate + water + carbon dioxide ⇌ calcium hydrogencarbonate

34.4 Uses of carbon dioxide

(*i*) In the manufacture of fizzy drinks
(*ii*) Solid carbon dioxide ('Drikold') is used as a refrigerant for food
(*iii*) Used to extinguish fires

34.5 Carbonates

Carbonates are salts of carbonic acid, H_2CO_3, formed by the complete replacement of hydrogen ions by metal ions. Table 34.1 summarises the more important properties of some common carbonates.

Table 34.1 Important properties of carbonates

Carbonate	*Formula*	*Colour*	*Solubility in water*	*Action of heat*	*Action of dilute acid*
Potassium	K_2CO_3	white	soluble	Not decomposed	
Sodium	Na_2CO_3	white	soluble		
Calcium	$CaCO_3$	white	insoluble		Any carbonate with any acid
Magnesium	$MgCO_3$	white	insoluble		liberates carbon dioxide. A salt
Zinc	$ZnCO_3$	white	insoluble	Decomposed to the oxide of	and water also formed
Iron	$FeCO_3$	light brown	insoluble	the metal and carbon dioxide	$MgCO_3(s) + H_2SO_4(aq) \rightarrow MgSO_4 + H_2O(l) + CO_2(g)$
Lead	$PbCO_3$	white	insoluble	$PbCO_3(s) \rightarrow PbO(s) + CO_2(g)$	
Copper	$CuCO_3$	bluish-green	insoluble		

N.B. Calcium carbonate and dilute sulphuric acid do not react because calcium sulphate is almost insoluble and forms a layer on the calcium carbonate preventing reaction.

The test for carbonate (see 41.2).

34.6 Sodium carbonate

Sodium carbonate can exist as the white anhydrous powder, Na_2CO_3, or in the hydrated form as sodium carbonate decahydrate, $Na_2CO_3.10H_2O$. These crystals are sometimes called washing soda and are efflorescent (see 11.7).

Sodium carbonate can be manufactured by the **Solvay process**. Ammoniacal brine (made by saturating a concentrated solution of sodium chloride with ammonia) descends a large tower called the carbonator in which there is an upward flow of carbon dioxide under pressure. Sodium hydrogencarbonate precipitates in the lower part of the tower.

$$NH_3(g) + CO_2(g) + H_2O(l) \rightarrow NH_4HCO_3(aq)$$
ammonia + carbon dioxide + water → ammonium hydrogencarbonate

$$NH_4HCO_3(aq) + NaCl(aq) \rightarrow NaHCO_3(s) + NH_4Cl(aq)$$
ammonium hydrogencarbonate + sodium chloride → sodium hydrogencarbonate + ammonium chloride

After filtration and washing to remove ammonium compounds, the sodium hydrogencarbonate is heated to convert it to sodium carbonate.

$$2NaHCO_3(s) \rightarrow Na_2CO_3(s) + H_2O(g) + CO_2(g)$$
sodium hydrogencarbonate → sodium carbonate + water + carbon dioxide

This process is efficient in that the raw materials (salt and limestone) are inexpensive and certain by-products can be re-cycled (see 38.5).

Sodium carbonate is used in the manufacture of glass, and in the softening of water (see 11.4).

34.7 Calcium carbonate

Calcium carbonate is found widely as limestone, chalk and marble. The deposits were formed from the shells of prehistoric sea creatures.

Although insoluble in water, it dissolves slowly in the presence of dissolved carbon dioxide to produce a solution of calcium hydrogencarbonate.

When calcium carbonate is heated in kilns to about 1000°C, it decomposes to form calcium oxide (quicklime).

$$CaCO_3(s) \rightarrow CaO(s) + CO_2(g)$$
calcium carbonate → calcium oxide + carbon dioxide

On adding water to the calcium oxide, it is converted (or slaked) to calcium hydroxide (slaked lime). This reaction is exothermic.

$$CaO(s) + H_2O(l) \rightarrow Ca(OH)_2(s)$$
calcium oxide + water → calcium hydroxide

34.8 Hydrogencarbonates

Hydrogencarbonates are acid salts of carbonic acid formed by the replacement of only one of the hydrogen ions by a metal ion.

Sodium hydrogencarbonate and potassium hydrogencarbonate exist as solids and calcium hydrogencarbonate exists only in solution. Any attempt to isolate solid calcium hydrogencarbonate causes it to decompose.

35 Polymers

35.1 Polymerisation

Polymerisation is the linking together of relatively small and simple molecules to form large units called **polymers**. The individual small molecules are called **monomers**. There are two types of polymerisation: **addition polymerisation** and **condensation polymerisation**.

35.2 Addition polymerisation

This process involves the addition of one substance to another to produce a single new substance. It may be represented as follows:

$$nM \rightarrow (M)_n$$

where M is a monomer unit

This type of polymerisation can only take place if the monomer is an unsaturated molecule (see 33.9). Many monomers, in fact, contain a double bond between two carbon atoms.

E.g. **Ethene** produces the polymer called **polyethene** (polythene) as shown in Fig. 35.1.

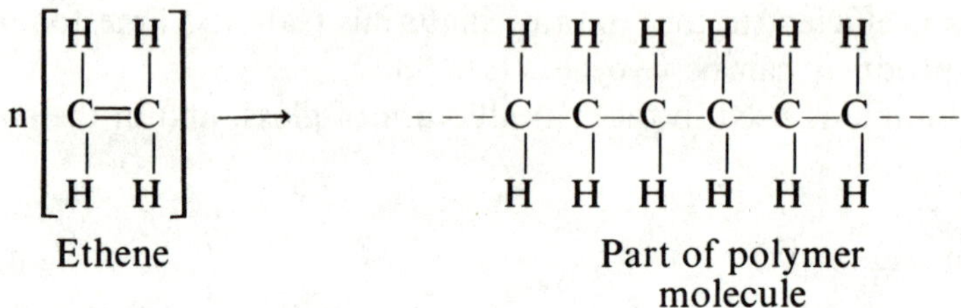

Fig. 35.1 Polymerisation of ethene

In the resulting polymer unit produced there are usually hundreds of ethene units combined together.

The polymer can be produced in two ways:

(*i*) Ethene gas is bubbled into a hydrocarbon solvent containing a complex catalyst. The temperature is below 100°C and the pressure of the gas is atmospheric. The polythene produced is called high density (HD) polythene.

(*ii*) Ethene is heated at high pressure (1000 atmospheres) and a temperature of 180°C with a little oxygen present as initiator. The polythene produced is called low density (LD) polythene.

Table 35.1 gives details of some common addition polymers and their uses.

35.3 Condensation polymerisation

When two molecules react together to form a larger molecule and eliminate a small molecule (*e.g.* water), this is called **condensation**. In condensation polymerisation, the monomer units are joined together with the elimination of small molecules.

Each monomer unit must contain **two reactive groups** otherwise no polymer is possible.

Nylon is an important man-made condensation polymer. Although there are various types of nylon, the commonest is nylon-6,6 (so-called because both starting materials contain six carbon atoms). For nylon-6,6, the starting materials are hexane-1,6-diamine and hexanedioic acid. They may be represented as follows:

H_2N—□—NH_2	HOOC—●—COOH
Hexane-1,6-diamine	Hexanedioic acid
(reactive group—NH_2)	(reactive group—COOH)

Table 35.1 Examples of addition polymers

Monomer	*Polymer*			
Formula/Name	*Name*	*Trade name*	*Formula*	*Uses*
$H_2C{=}CH_2$ ETHENE	Polyethene	Polythene	$\left(-CH_2-CH_2-\right)_n$	Plastic sheets, pipes, plastic bags
$H_2C{=}CH(CH_3)$ PROPENE	Polypropene	Propathene	$\left(-CH_2-CH(CH_3)-\right)_n$	Plastic sheets, electric insulators, washing up bowls
$H_2C{=}CHCl$ VINYL CHLORIDE (CHLOROETHENE)	Polyvinylchloride	PVC	$\left(-CH_2-CHCl-\right)_n$	Records, clothes, electrical wire insulators
$H_2C{=}CH(C_6H_5)$ STYRENE (PHENYLETHENE)	Polystyrene	—	$\left(-CH_2-CH(C_6H_5)-\right)_n$	Packaging materials, ceiling tiles, plastic model kits
$H_2C{=}CH(COOCH_3)$ METHYL METHACRYLATE	Polymethyl-methacrylate	Perspex	$\left(-CH_2-CH(COOCH_3)-\right)_n$	Substitute for glass
$H_2C{=}CH(CN)$ ACRYLONITRILE	Polyacry-lonitrile	Orlon, Courtelle, Acrilan	$\left(-CH_2-CH(CN)-\right)_n$	Synthetic fibre
$F_2C{=}CF_2$ TETRAFLUOROETHENE	Polytetra-fluoroethene	Teflon PTFE	$\left(-CF_2-CF_2-\right)_n$	Coating for non-stick saucepans; bridge bearings

One of the reactive $-NH_2$ groups on the hexane-1,6-diamine molecule reacts with one of the reactive $-COOH$ groups on the hexanedioic acid molecule with the elimination of a molecule of water. The product still contains two reactive groups and a series of similar reactions take place resulting in the formation of a polymer (Fig. 35.2).

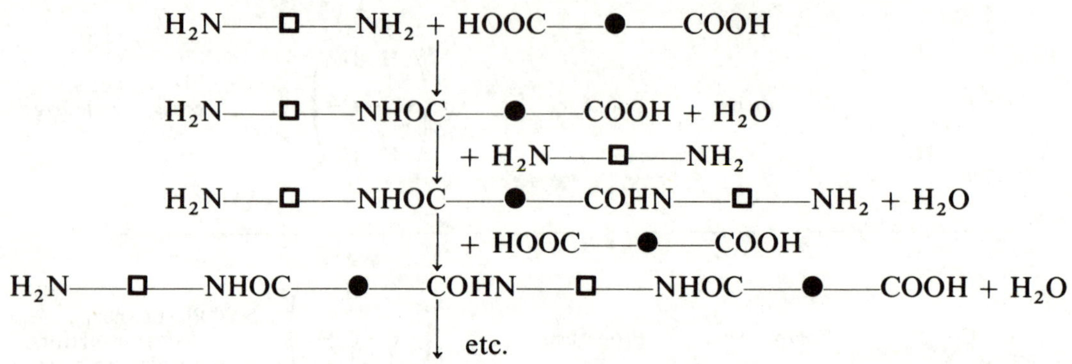

Fig. 35.2 Fermentation of nylon-6,6 (a condensation polymer)

35.4 Thermosetting and thermoplastic polymers

Bakelite is also a very common condensation polymer and is used in the manufacture of electrical fittings, pens, telephones, *etc.* It is an example of a **thermosetting plastic**, *i.e.* it can be moulded into the required shape whilst hot but becomes hard and brittle when cool. It cannot be melted again once it has solidified.

In contrast, plastics such as polythene may be alternately softened and hardened and are known as **thermoplastic polymers**.

35.5 Polymers from natural sources

Many substances which occur in living organisms are condensation polymers.

Proteins are long chain condensation polymers which resemble nylon in structure. They are formed by the polymerisation of a limited number of amino acids, *e.g.*

$$\begin{array}{c} H \\ | \\ H_2N-C-COOH \\ | \\ R \end{array}$$

Polysaccharides are condensation polymers.

36 Fuels

36.1 Introduction

A fuel is a substance which can be used to produce energy. Most fuels, and certainly the ones considered in this section, liberate energy when burnt in air or oxygen. These fuels are compounds of carbon and hydrogen, the elements carbon and hydrogen and carbon monoxide.

Fuels can be classified as solid fuels, liquid fuels and gaseous fuels.

36.2 Solid fuels

The most important solid fuel is coal. Coal is a fossil fuel composed largely of carbon. A typical coal sample contains 80% carbon, 6% hydrogen, 6% nitrogen, 5% oxygen and 3% ash.

The coal was formed by the action of heat and pressure on plants over millions of years. Anthracite is a form of coal consisting of a larger percentage of carbon resulting from a more complete decomposition of the plants.

When coal is burnt in excess air, carbon dioxide gas is produced. However, other products formed include water vapour and sulphur dioxide. The sulphur dioxide produced causes pollution problems (see 38.6), and for this reason smokeless fuel is preferred to coal.

Destructive distillation of coal

When coal is heated to about 1000°C in the absence of air, it does not burn. It splits up and forms four products—coal gas, ammonia solution, coal tar and coke (or smokeless fuel). Suitable apparatus is shown in Fig. 36.1.

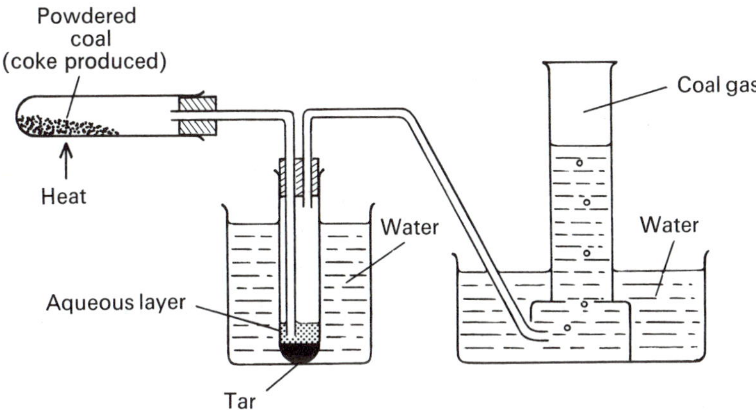

Fig. 36.1 Destructive distillation of coal

The coal gas produced used to be used for household gas supply but it has now been replaced by natural gas. Coal gas consists chiefly of hydrogen with smaller amounts of methane and carbon monoxide.

The ammonia produced collects in the water. It can be used to make fertilisers. Coal tar is a complicated mixture of hydrocarbons and closely related compounds. It can be separated by fractional distillation into valuable chemicals, *e.g.* benzene, toluene.

36.3 Liquid fuels

Liquid fuels include petrol and paraffin. They are often obtained from crude oil (see 33.20). Petrol is a mixture of alkanes including octane. Compounds including tetraethyl lead $Pb(C_2H_5)_4$ are added to the petrol to improve its combustion properties.

Other liquid fuels include the family of alcohols (see 33.22) including methanol and ethanol.

36.4 Gaseous fuels

The most important gaseous fuel is methane (CH_4). It is found as natural gas, either alone or associated with crude oil. Natural gas and oil were formed by the action of heat and pressure on the remains of small sea creatures. They are trapped below ground by layers of rock that do not allow the natural gas to escape (Fig. 36.2).

Other gaseous fuels include **water gas** and **producer gas**.

Producer gas is formed when air is passed over red-hot coke (carbon).

$$2C(s) + O_2(g) + N_2(g) \rightarrow 2CO(g) + N_2(g)$$
$$\text{carbon} + \text{air} \rightarrow \text{carbon monoxide} + \text{nitrogen}$$

This reaction is exothermic and the producer gas yielded is a mixture of carbon monoxide and nitrogen. Producer gas is not a good fuel because it contains a large proportion of nitrogen which does not burn. It is, however, cheap to produce.

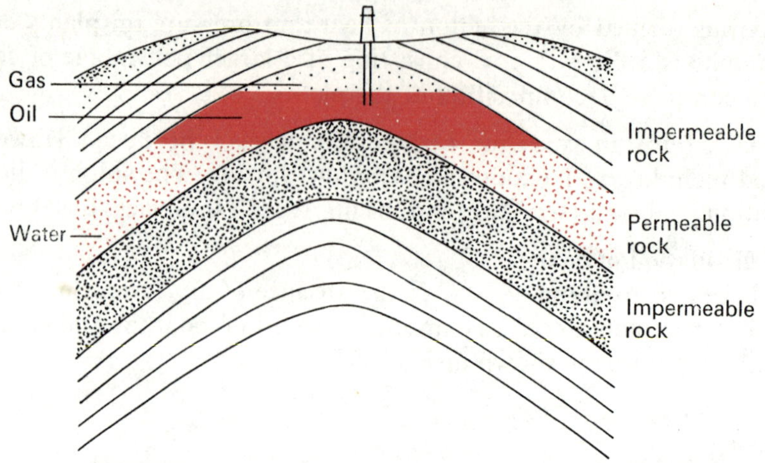

Fig. 36.2 Formation of oil and natural gas

Water gas is produced when steam is passed over white-hot coke (carbon).

$$C(s) + H_2O(g) \rightleftharpoons CO(g) + H_2(g)$$
carbon + water ⇌ carbon monoxide + hydrogen

This reaction is endothermic and the coke has to be reheated if the reaction is to continue. Water gas is a mixture of hydrogen and carbon monoxide. It is a good gaseous fuel because both carbon monoxide and hydrogen burn.

36.5 Which is the best fuel?

There are a number of factors that determine which fuel is best for a particular purpose. These include:

(*i*) price;
(*ii*) ease of storage;
(*iii*) energy value of fuel;
(*iv*) ease of transport;
(*v*) ease of ignition;
(*vi*) percentage of impurities;
(*vii*) pollution products.

37 Energy in chemistry

37.1 Introduction

A consideration of energy, and more particularly energy change, is a fundamental aspect of a chemistry course. In this section the distinction between Physics and Chemistry becomes blurred.

Energy can be defined as the capacity to cause a change in the condition of the surroundings. There are different forms of energy. The most important forms of energy to a chemist are heat energy and electrical energy.

Energy can neither be created nor destroyed in any process. It is just changed from one form to another. This is called the **Law of Conservation of Energy,** which is related to the Law of Conservation of Mass. The work of Einstein has shown that mass and energy can be interconverted.

37.2 Energy possessed by chemicals

All substances are composed of small particles. At any temperature above absolute zero (−273°C), the particles are in motion. These particles can possess two types of energy:

(i) Kinetic energy. Particles that are moving possess kinetic energy (energy of movement). As the temperature rises, the particles move faster and possess more kinetic energy.

(ii) Chemical energy (or bonding energy). When bonds are formed energy is released. If the bonds holding particles together are to be broken, energy has to be supplied.

It is not possible to measure the total energy possessed by any chemical in the laboratory. It is only possible to measure energy changes.

The energy change that accompanies a chemical reaction is due to changes in chemical or bonding energy between reactant and product.

37.3 Exothermic processes

There are many examples where energy is liberated during the reaction and the temperature of the system rises.

For example, the burning of carbon produces energy. Such a reaction is called an exothermic reaction. The quantity of energy produced depends on the quantity of carbon burnt.

$$C(s) + O_2(g) \rightarrow CO_2(g)$$
carbon + oxygen → carbon dioxide ($\Delta H = -393.5$ kJ)

This information tells a chemist that if 1 mole of carbon atoms (12 g) is burnt in an adequate supply of oxygen, 393.5 kJ of energy are liberated. ΔH is called the enthalpy (or heat) of reaction and is **negative** if the reaction is **exothermic**.

This process can be summarised in Fig. 37.1.

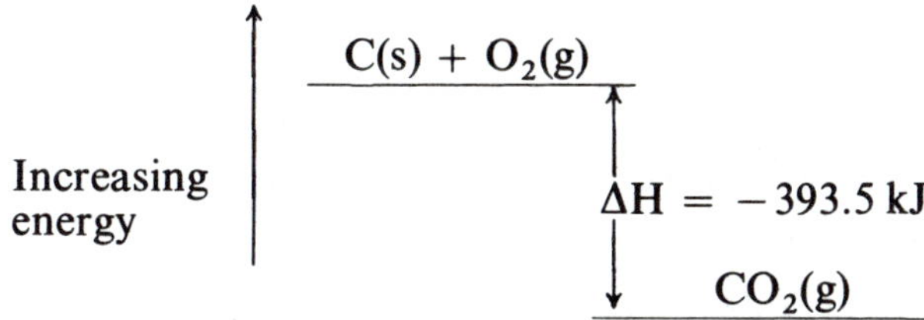

Fig. 37.1 Energy diagram for the complete combustion of carbon

During an exothermic reaction, more energy is produced from the formation of new bonds than the energy required to break existing bonds, and therefore there is a surplus of energy which is lost to the surroundings.

37.4 Endothermic processes

There are few reactions where energy is absorbed from the surroundings during the reaction and the temperature falls.

For example, the formation of hydrogen iodide from hydrogen and iodine is an endothermic reaction.

$$H_2(g) + I_2(g) \rightleftharpoons 2HI(g)$$
hydrogen + iodine ⇌ hydrogen iodide ($\Delta H = +52$ kJ)

ΔH is **positive**, in this case, because the reaction is **endothermic**. This process is summarised in Fig. 37.2.

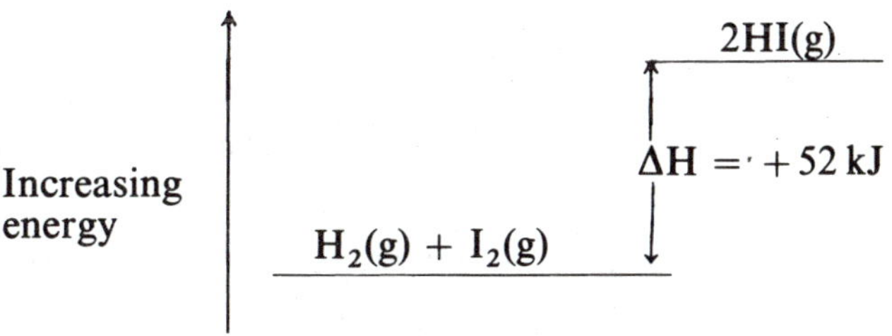

Fig 37.2 Energy diagram for the synthesis of hydrogen iodide

During an endothermic reaction, more energy is required to break bonds than is liberated when new bonds are formed. This deficiency in energy has to be made up from the surroundings.

37.5 Activation energy

If the products contain less energy than the reactants, it might be expected that the exothermic changes would always occur spontaneously. This is not the case. Before the reaction can take place energy has to be supplied, called the activation energy, to start the reaction off (Fig. 37.3).

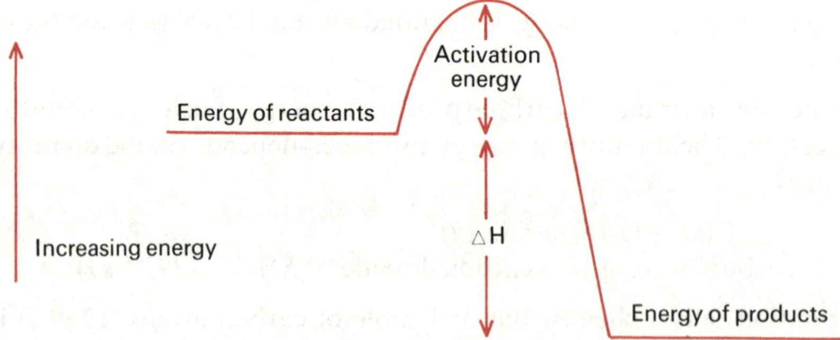

Fig. 37.3 Energy changes during a reaction

The function of a catalyst is usually to speed up a reaction by lowering the activation energy barrier (see 20.4).

37.6 Fundamental definitions

Heat of formation. The heat liberated or absorbed when 1 mole of a substance is formed from its constituent elements.

Heat of combustion. The heat liberated when 1 mole of a substance is completely burnt in excess oxygen.

Heat of neutralisation. The heat is liberated when 1 mole of hydrogen ions $H^+(aq)$ reacts with 1 mole of hydroxide ions $OH^-(aq)$.

$$H^+(aq) + OH^-(aq) \rightarrow H_2O(l)$$

The heat of neutralisation of many acids and alkalis is -58 kJ/mole.

37.7 To find the heat of combustion of an alcohol

A small spirit lamp containing the alcohol is weighed accurately. During the weighing the spirit lamp should be covered to prevent evaporation of the alcohol.

A known mass of water is taken in a metal can. The temperature of the water is recorded. The spirit lamp is lit and placed underneath the can. The can is heated until the temperature has risen about 30°C. The final temperature is recorded. The spirit lamp is then reweighed to find the mass of alcohol burnt.

Sample results:

Mass of spirit lamp + ethanol before burning = 21.94 g
Mass of spirit lamp + ethanol after burning = 21.10 g
Mass of ethanol burnt = 0.84 g
Mass of water in the can = 100 g

(You can ignore the heat required to raise the temperature of the can.)

Specific heat capacity of the water = 4.2 kJ/kg/°C
Temperature rise = 30°C

$$\text{Heat gained by water} = \text{mass (in kg)} \times \text{temp. rise} \times \text{specific heat capacity}$$
$$= \frac{1}{10} \times 30 \times 4.2\,\text{kJ}$$
$$= 12.6\,\text{kJ}$$

It is assumed that all the heat produced when the ethanol burns is used to heat up the water.

12.6 kJ is produced when 0.84 g of ethanol burns

$\frac{12.6}{0.84}$ kJ is produced when 1 g of ethanol burns

Mass of 1 mole of ethanol molecules C_2H_5OH = 46 g

Heat produced when 46 g of ethanol burns $= 46 \times \frac{12.6}{0.84}$ kJ

Since the combustion of ethanol is exothermic, the heat of combustion

$$\Delta H = -690\,\text{kJ/mole}$$

However, the heat of combustion of ethanol in the data book is −1370 kJ/mole.

The big difference between the theoretical and practical heats of combustion is due to the heat lost to the surroundings.

37.8 Cells and batteries

The energy produced during a chemical reaction can be used in various ways. These include:

(*i*) heat (chemical energy → heat energy);
(*ii*) to do work (chemical energy → mechanical energy);
(*iii*) to produce electricity (chemical energy → electrical energy).

Not all the energy produced, ΔH, can be converted to electricity. The maximum amount of energy that can be converted to electricity is ΔG. The rest of the energy is wasted.

A simple cell is produced when rods of zinc and copper are dipped into a solution of copper(II) sulphate.

The reaction taking place is

$$Zn(s) + CuSO_4(aq) \rightarrow ZnSO_4(aq) + Cu(s)$$
$$\text{zinc} + \text{copper(II) sulphate} \rightarrow \text{zinc sulphate} + \text{copper}$$

A simple cell does not operate well due to polarisation. The polarisation can be prevented in various ways. One possible way of setting up a simple cell is shown in Fig. 37.4.

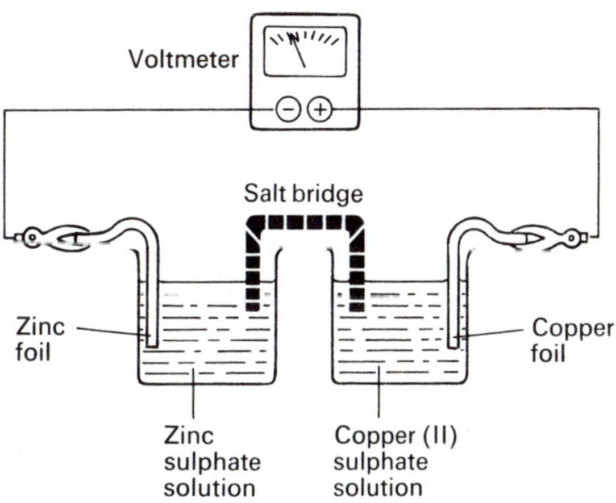

Fig. 37.4 A simple cell

37.9 FUEL CELLS

A fuel cell is an efficient means of converting chemical energy into electrical energy. A continuous supply of electrical energy can be obtained if there is a continuous supply of fuel.

In a hydrogen–oxygen fuel cell, hydrogen and oxygen combine to form water. The energy is released as electricity.

Fuel cells have been used in vehicles and space rockets. They produce electricity without pollution.

---

38 Social and economic considerations

38.1 INTRODUCTION

It is a modern tendency to include social and economic considerations into Chemistry courses. It is never, however, a whole question on an examination paper but is incorporated in existing questions. The questions that are suitable for consideration of the social and economic aspects are usually industrial processes. The most important are:

Extraction of iron (see 15.7)
Contact process (see 25.2)
Haber process (see 27.3)
Extraction of aluminium (see 15.6)
Nitric acid manufacture (see 28.3)
Sodium hydroxide manufacture (see 14.3)
Extraction of sodium (see 15.5)
Sodium carbonate manufacture (see 34.6)
Hydrogen manufacture (see 10.4)
Oil refining (see 33.20)

Having been through this section, refer back to each process to see how the principles apply. There are five principles to be considered.

38.2 CHOICE OF SITE FOR THE PROCESS

There are a number of factors which will determine the best place to site a new factory. If bulky raw materials are required it is often advisable to site the factory near the natural deposits or on the coast if the raw materials are to be imported. For this reason, oil refineries are often sited on the coast. If more than one raw material is required, then it is necessary to decide which raw material will be more difficult to transport.

If a process requires a large amount of energy, *e.g.* extraction of aluminium, it may be advisable to site the factory close to hydroelectric power for cheap electricity.

Another consideration when choosing a site is the location of the customers for the product. The building of a sodium hydroxide factory in north Cheshire (close to salt deposits) can be justified by processes nearby (*e.g.* soap making), which uses large quantities of sodium hydroxide. Another example is the siting of a factory producing oxygen by fractional distillation of liquid air adjacent to a factory producing electric light bulbs. The inert gases produced during the fractional distillation being used to fill electric light bulbs.

A suitable supply of labour and good communications (road, rail *etc.*) are obviously other considerations.

If a process requires a supply of water for cooling purposes, the factory needs to be sited near a lake or river.

If the process uses large quantities of toxic or inflammable materials or operates at a high pressure, it would be advisable to site the factory away from housing areas (see 42).

38.3 Using the minimum amount of energy

In order to reduce energy costs, it is necessary to consider ways of re-using energy or reducing working temperatures. In the extraction of sodium from sodium chloride, the melting point of sodium chloride is reduced from about 800°C to about 550°C by adding an impurity of calcium chloride. This does not affect the products but reduces the fuel required to keep the sodium chloride molten.

In the Contact process, reducing the temperatures of the catalyst chamber saves energy but also causes a greater conversion of sulphur dioxide to sulphur trioxide. It also, however, slows down the process. The slowing down of the process when temperature is reduced has always to be remembered.

Often, following an exothermic reaction, the products are at a higher temperature than is required. Under these circumstances, a heat exchanger can be used to remove heat from the products to be re-used elsewhere. For example in the first stage of the conversion of ammonia and air to nitric acid, the reaction is exothermic and the temperature of the gases has to be reduced considerably. A heat exchanger or a series of heat exchangers can be used to transfer the energy to some other part of the process. The energy possessed by hot gases, that are being produced and expelled through a chimney, is removed by heat exchangers. Often waste gases contain carbon monoxide and other gases that can be burnt to produce energy rather than being allowed to escape into the atmosphere.

38.4 Continuity of a process

A process that works continuously rather than in batches is usually more economic. A blast furnace producing iron continuously can produce 1800 tonnes of iron per day. It can work continuously because the impurities in the iron are being removed in the slag rather than accumulating in the furnace.

Often a process that appears to be continuous is not in practice. For example, the manufacture of nitric acid would appear to be continuous but a run of production of more than a few days is unusual because the platinum catalyst gauzes have to be replaced. In order to do this, the plant has to be shut down.

38.5 Re-use of by-products or production of valuable by-products

The economics of a process depends heavily upon the usefulness of the by-products of the process. In the manufacture of sodium hydroxide by electrolysis, the by-products (chlorine and hydrogen) are very valuable and can be easily sold.

When ammonia is being produced industrially from nitrogen and hydrogen, only about 10% of the gases are converted to ammonia. The unused nitrogen and hydrogen are then re-cycled.

38.6 Pollution considerations

The control of waste from a factory is monitored by inspectors. It is necessary to treat wastes correctly before they are allowed to escape into the air or into water.

One of the major impurities in waste gases, which needs to be removed, is sulphur dioxide. When this gets into the air it becomes oxidised to sulphuric acid. In addition to polluting the atmosphere, the escape of sulphur dioxide is wasting valuable sulphur that could be re-used.

It is necessary to treat waste water to remove toxic materials like cyanides and heavy metals. It is usual to precipitate these substances to facilitate their removal. Apart from these impurities, it is necessary to monitor the temperature, pH and nitrogen content of the water. If ammonia is present in the water, bacteria will oxidise the ammonia to nitrates and remove oxygen from the water. This is called **eutrophication** and can be recognised by a green algae on the water. This green algae prevents light entering water, and further oxygen is removed from the water when underwater vegetation dies and decays.

39 Radioactivity

39.1 Introduction

The nuclei of some heavier atoms are unstable and tend to split up with the emission of certain types of radiation and the formation of new elements. This decay is called **radioactive decay.**

It is possible for lighter atoms, containing a large proportion of neutrons, to undergo radioactive decay, *e.g.* $^{3}_{1}H$ is a radioactive isotope of hydrogen.

39.2 Types of radiation emitted

The radiation emitted can be of three types:

α-particles. These are positively charged particles each of which is identical with the nucleus of a helium atom, *i.e.* two protons and two neutrons but no electrons. They are comparatively heavy and slow moving and have little penetrating power. For example, they are unable to penetrate a piece of paper. They are deflected by magnetic and electrostatic fields. Because of their low penetrating power they cannot be detected by the usual apparatus used in the laboratory.

β-particles. These are negatively charged particles. They are in fact electrons. They have greater penetrating power and are deflected by magnetic and electric fields. They are detected by the apparatus used in the laboratory.

γ-rays. These are high energy electromagnetic waves with zero charge. They are very penetrating and are unaffected by electric and magnetic fields. Although they penetrate the apparatus, they are not recorded on laboratory apparatus.

39.3 Apparatus used for radioactivity measurements

In experiments in the laboratory, radioactivity measurements are made using a **Geiger counter** attached to a suitable counting tube. The sample is placed in the counting tube and the measurements are made on the Geiger counter.

Because radioactive decay is taking place in the atmosphere and surroundings, it is necessary to correct for **background radiation**. In an experiment, if the reading on the Geiger counter is 512 counts per second (cps) when the sample is in place but 10 cps in the absence of any radioactive sample, the corrected reading would be 502 cps (*i.e.* 512 – 10).

39.4 Half life $t_{1/2}$

The rate of decay of a radioactive isotope is independent of temperature. The time taken for half the mass of a radioactive isotope to decay is called the **half life** and is a characteristic of the isotope. It is the time taken for the corrected reading on the Geiger counter to fall to half of its original value.

$$\text{Half life of } ^{214}_{84}\text{Po} = 1.5 \times 10^{-4}\text{ s}$$
$$\text{Half life of } ^{226}_{88}\text{Ra} = 1620\text{ years}$$

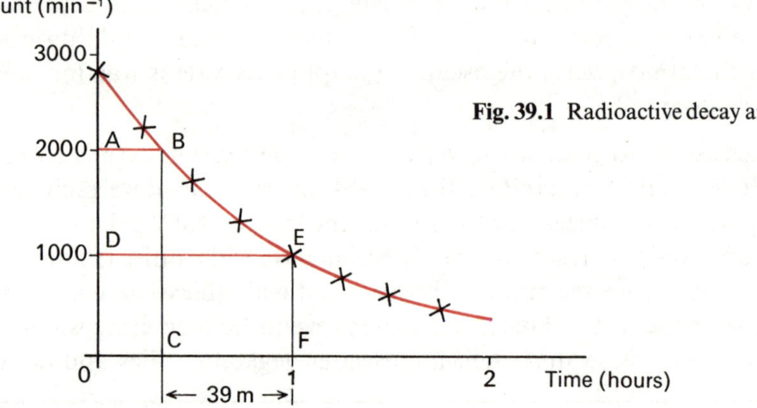

Fig. 39.1 Radioactive decay and half life

A graph can be plotted as in Fig. 39.1 if regular readings are taken at intervals and then corrected. If a convenient reading is taken (say 2000) and the lines AB and BC drawn, then the lines DE and EF are drawn at a reading which is half of the original reading. The time difference between BC and EF is called the half life.

39.5 Changes that accompany radioactive decay

Radioactive decay can change the number of protons and neutrons in the nucleus of an atom.

When an isotope loses an α-particle, it loses two protons and two neutrons. For example, α-decay of uranium-238.

$$^{238}_{92}U \rightarrow ^{4}_{2}He + ^{234}_{90}Th$$
$$\text{Uranium-238} \rightarrow \alpha\text{-particle} + \text{Thorium-234}$$

(N.B. In $^{238}_{92}U$, 238 is the mass number and 92 is the atomic number.)

The product of α-decay contains two fewer protons and, therefore, is two places to the left of uranium in the Periodic Table.

When an isotope undergoes β-decay, a neutron in the nucleus changes to a proton and an electron, which is emitted. The product, therefore, has one more proton than the starting isotope, and is therefore one place to the right in the Periodic Table. Thorium-234 undergoes β-decay.

$$^{234}_{90}Th \rightarrow ^{234}_{91}Pa + \text{electron}$$
$$\text{Thorium-234} \rightarrow \text{Protactinium-234} + \text{electron}$$

The emission of γ-rays does not change the number of protons or neutrons in the isotope. γ-radiation usually accompanies other types of emission.

39.6 Uses of radioactivity

Radioactivity has a large number of uses in industry. These include:

(*i*) treating cancer by subjecting a patient to controlled amounts of γ-radiation from a cobalt-60 source;
(*ii*) sterilising instruments and equipment using γ-radiation;
(*iii*) controlling the thickness of paper, rubber, metals and plastic accurately;
(*iv*) controlling the filling of packets and containers;
(*v*) tracing the course of a process by following a radioactive isotope;
(*vi*) The energy produced by radioactive fission of uranium-235 is used within a nuclear power station to provide electricity.

The uses of a radioactive isotope depend upon its half life. If the half life is too short or too long it can have little practical application.

40 Quantitative volumetric chemistry

40.1 Introduction

In this section we are concerned with the volumes of standard solutions that react exactly together; with this information various calculations can be done.

In unit 12, it was explained that the concentration of a solution can be expressed in terms of molarity.

E.g. What is the concentration of a solution of hydrochloric acid containing 7.3 g of hydrogen chloride in 100 cm³ of solution? (H = 1, Cl = 35.5)

$$\text{Mass of 1 mole of hydrogen chloride HCl} = 1 + 35.5\text{ g} = 36.5\text{ g}$$

7.3 g of hydrogen chloride in 100 cm³ of solution has the same molarity as 73 g of hydrogen chloride in 1000 cm³ of solution.

$$\text{The solution contains } \frac{73}{36.5} \text{ moles of hydrogen chloride}$$

$\therefore$ the solution is 2 M.

Remember: 1 mole of a substance dissolved to make 1000 cm³ of solution produces a molar (M) solution.

A solution of known concentration or known molarity is called a **standard solution**. In volumetric chemistry, a series of **titrations** are carried out. In each titration, a solution A is added in small measured quantities, from a burette, to a fixed volume of a solution B, measured with a pipette, in the presence of an indicator. At least one of the solutions must be a standard solution. The addition of a solution A is continued until the indicator just changes colour. At this stage, called the **end point**, the two substances in solution are present in quantities that exactly react.

In any titration, the accuracy of the student is very important.

40.2 TITRATION OF SODIUM HYDROXIDE WITH STANDARD SULPHURIC ACID

A solution of sodium hydroxide (of unknown concentration) is going to be standardised (*i.e.* its concentration found) by titration with 0.1 M sulphuric acid.

Exactly 25 cm³ of sodium hydroxide solution is added to a conical flask using a pipette. A couple of drops of screened methyl orange (indicator) is added and the solution turns green. (Litmus is not sufficiently sensitive.) 0.1 M sulphuric acid is put into the burette and the reading on the burette recorded. The sulphuric acid is added to the flask in small volumes. After each addition, the flask is swirled. The process is continued until the solution turns colourless. The final reading on the burette is recorded.

The procedure is repeated until consistent results are obtained. Sample results:

Volume of sodium hydroxide solution = 25.00 cm³
Volume of sulphuric acid = 24.00 cm³

(This result would be the average of the results obtained.)

Molarity of sulphuric acid = 0.1 M

The reaction is represented by the equation:

$$2NaOH(aq) + H_2SO_4(aq) \rightarrow Na_2SO_4(aq) + 2H_2O(l)$$

sodium hydroxide + sulphuric acid → sodium sulphate + water

If 1000 cm³ of M sulphuric acid were used, this would contain 1 mole of sulphuric acid.

$$\therefore 24.00\text{ cm}^3\ 0.1\text{ M sulphuric acid contains } 0.1 \times \frac{24}{1000} \text{ moles of sulphuric acid} = 0.0024 \text{ moles sulphuric acid}$$

From the equation:

1 mole of sulphuric acid exactly reacts with 2 moles sodium hydroxide

$\therefore$ 0.0024 moles of sulphuric acid exactly reacts with 0.0048 moles sodium hydroxide.

Now, 0.0048 moles of sodium hydroxide is contained in 25.00 cm^3 of solution.

∴ 1000 cm^3 of sodium hydroxide solution would contain

$$\frac{0.0048 \times 1000}{25} = 0.192 \text{ moles}$$

Molarity of sodium hydroxide = 0.192 M

40.3 Titration of standard sodium carbonate with hydrochloric acid

A solution of hydrochloric acid (of unknown concentration) is going to be standardised by titration with 0.1 M sodium carbonate solution.

Exactly 25 cm^3 of sodium carbonate solution is added to a conical flask from a pipette. A couple of drops of screened methyl orange is added. Hydrochloric acid is added to the flask in small volumes and the volume of acid required to change the colour of the indicator is found. Again further titrations are carried out until consistent results are obtained.

Sample results:

Volume of sodium carbonate solution = 25.00 cm^3
Volume of hydrochloric acid = 23.50 cm^3 (average)
Molarity of sodium carbonate solution = 0.1 M

The equation for the reaction is:

$$Na_2CO_3(aq) + 2HCl(aq) \rightarrow 2NaCl(aq) + CO_2(g) + H_2O(l)$$
sodium carbonate + hydrochloric acid → sodium chloride + carbon dioxide + water

1000 cm^3 of M sodium carbonate contains 1 mole of sodium carbonate

∴ 25 cm^3 of 0.1 M sodium carbonate contains

$$\frac{0.1 \times 25}{1000} \text{ moles sodium carbonate} = 0.0025 \text{ moles sodium carbonate}$$

From the equation:

1 mole of sodium carbonate reacts with 2 moles of hydrochloric acid; 0.0025 moles of sodium carbonate reacts with 0.005 moles of hydrochloric acid. Now, 0.005 moles of hydrochloric acid is contained in 23.50 cm^3 of solution.

$$\therefore \text{ 1000 cm}^3 \text{ of solution would contain } \frac{0.005 \times 1000}{23.5} \text{ moles} = 0.213 \text{ moles}$$

Molarity of hydrochloric acid = 0.213 M

N.B. If phenolphthalein had been used as indicator, the volume of acid required would have been 11.75 cm^3 (exactly half the volume required when screened methyl orange was used). This is because phenolphthalein is detecting the end point in the reaction.

$$Na_2CO_3(aq) + HCl(aq) \rightarrow NaHCO_3(aq) + NaCl(aq)$$
sodium carbonate + hydrochloric acid → sodium hydrogencarbonate + sodium chloride

40.4 Points to remember when doing volumetric experiments

Simple volumetric experiments are frequently set for practical examinations. When doing this type of experiment accuracy is very important and the following points should be remembered:

(*i*) Shake up all solutions thoroughly before use to make sure the solution is the same throughout.
(*ii*) Rinse out the conical flask with distilled water only.
(*iii*) The burette and pipette should be rinsed out with the solution that is to go into them. These rinsing solutions should then be discarded.

(*iv*) The last drop of solution in the pipette should not be blown or shaken out of the pipette.
(*v*) Distilled water can be added to the conical flask during the titration.

Suppose an exactly 0.1 M solution of a metal hydroxide $M(OH)_x$ is provided together with 0.2 M hydrochloric acid. It is possible to find the value of x and then the equation from a volumetric experiment.

E.g. 25.00 cm³ of 0.1 M metal hydroxide exactly reacts with 25.00 cm³ of 0.2 M hydrochloric acid.

25.00 cm³ of 0.1 M metal hydroxide contains

$$\frac{0.1 \times 25}{1000} \text{ moles of metal hydroxide} = \frac{1}{400} \text{ mole} = 0.0025 \text{ moles}$$

25.00 cm³ of 0.2 M hydrochloric acid contains

$$\frac{0.2 \times 25}{1000} \text{ moles of hydrochloric acid} = \frac{1}{200} \text{ moles} = 0.005 \text{ moles}$$

$\therefore$ 1 mole of $M(OH)_x$ would exactly react with 2 moles of hydrochloric acid.

So the formula of the metal hydroxide is $M(OH)_2$ and the equation is

$$M(OH)_2(aq) + 2HCl(aq) \rightarrow MCl_2(aq) + 2H_2O(l).$$

40.6 Titration without an indicator

The end point in a titration can be found by following the pH or electrical conductivity during the reaction. For example, the titration of barium hydroxide solution with dilute sulphuric acid can be carried out using electrical conductivity to detect the end point. At the end point, the electrical conductivity is zero.

$$Ba(OH)_2(aq) + H_2SO_4(aq) \rightarrow BaSO_4(s) + 2H_2O(l)$$
barium hydroxide + sulphuric acid → barium sulphate + water

Figure 40.1 shows the apparatus required and the graph obtained. The end point is the minimum point on the graph.

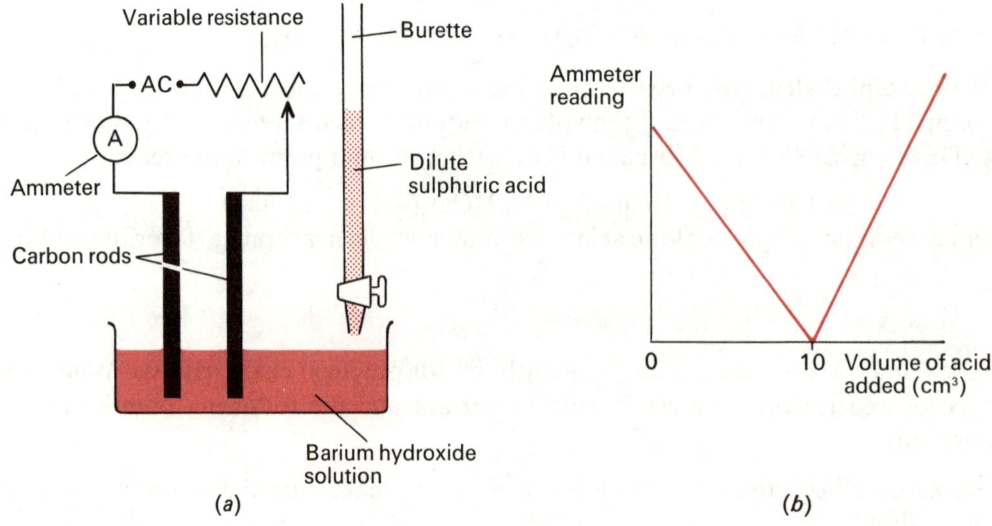

Fig. 40.1 Neutralisation of barium hydroxide solution with dilute sulphuric acid: (*a*) apparatus to follow the reaction by measuring conductivity changes; (*b*) graph of sample results

40.7 To find the number of molecules of water of crystallisation in a sample of sodium carbonate (see 11.6)

Hydrated sodium carbonate has a formula $Na_2CO_3.yH_2O$. In this experiment the value of y is going to be calculated.

7.15 g of hydrated sodium carbonate was dissolved to make 250 cm³ of solution. A 25 cm³ sample of this solution was taken and screened methyl orange indicator added. 50.00 cm³ of 0.1 M hydrochloric acid was added to change the colour of the indicator.

$$Na_2CO_3(aq) + 2HCl(aq) \rightarrow 2NaCl(aq) + H_2O(l) + CO_2(g)$$

sodium carbonate + hydrochloric acid → sodium chloride + water + carbon dioxide

50.00 cm³ of 0.1 M hydrochloric acid contains

$$\frac{0.1 \times 50}{1000} \text{ moles of hydrogen chloride} = 0.005 \text{ moles}$$

0.005 moles of hydrochloric acid react with 0.0025 moles of sodium carbonate.

Let the molarity of the sodium carbonate solution be zM.
1000 cm³ of M sodium carbonate contains 1 mole of sodium carbonate.

25 cm³ of M sodium carbonate contains $\frac{1 \times 25}{1000}$ moles of sodium carbonate.

25 cm³ of zM sodium carbonate contains $\frac{1 \times 25 \times z}{1000}$ moles of sodium carbonate.

Now,
$$\frac{1 \times 25 \times z}{1000} = 0.0025$$
$$z = 0.1$$

Sodium carbonate solution is 0.1 M.

Concentration of Na_2CO_3 in sodium carbonate solution $= 0.1 \times 106$
$= 10.6$ g/dm³

Concentration of $Na_2CO_3.yH_2O = 7.15 \times 4 = 28.6$ g/dm³

∴ 28.6 g of sodium carbonate crystals contain 10.6 g of sodium carbonate Na_2CO_3 and 18.0 g of water. ($Na_2CO_3 = 106$, $H_2O = 18$)

0.1 moles of sodium carbonate combines with 1 mole of water.

∴ 1 mole of sodium carbonate combines with 10 moles of water.

$$\therefore y = 10$$

Sodium carbonate crystals are $Na_2CO_3.10H_2O$.

41 Qualitative chemistry

41.1 Introduction

In this section the tests that can be used to identify substances are explained. You are advised to check in your examination board syllabus to find out which tests are included. For CSE only potassium, sodium, calcium, copper(II), iron(II), iron(III), zinc, lead, ammonium, chloride, nitrate, sulphate and carbonate *are usually required.*

41.2 Tests for anions (negative ions)

Carbonate (CO_3^{2-})
When dilute hydrochloric acid is added to a carbonate, carbon dioxide gas is produced. No heat is required. The carbon dioxide turns limewater milky.

E.g. $$Na_2CO_3(s) + 2HCl(aq) \rightarrow 2NaCl(aq) + H_2O(l) + CO_2(g)$$
sodium carbonate + hydrochloric acid → sodium chloride + water + carbon dioxide

Hydrogencarbonate (HCO_3^-)

Hydrogencarbonates behave in a similar way to carbonates with dilute hydrochloric acid.

When a solution of a hydrogencarbonate is heated, carbon dioxide is produced.

E.g. $$2NaHCO_3(aq) \rightarrow Na_2CO_3(aq) + H_2O(l) + CO_2(g)$$
sodium hydrogencarbonate → sodium carbonate + water + carbon dioxide

Nitrite (NO_2^-)

When dilute hydrochloric acid is added to a nitrite, brown nitrogen dioxide gas is produced and the solution turns pale blue. No heat is required. The nitrogen dioxide turns blue litmus paper red but does not bleach it.

Sulphide (S^{2-})

When dilute hydrochloric acid is added to a sulphide, colourless hydrogen sulphide gas is produced. The hydrogen sulphide smells of bad eggs and turns filter paper, soaked in lead nitrate solution, black.

E.g. $$Na_2S(s) + 2HCl(aq) \rightarrow 2NaCl(aq) + H_2S(g)$$
sodium sulphide + hydrochloric acid → sodium chloride + hydrogen sulphide

Sulphite (SO_3^{2-})

When dilute hydrochloric acid is added to a sulphite and the mixture heated, colourless sulphur dioxide gas is produced. The sulphur dioxide has a pungent odour and turns potassium dichromate from orange to green. It does not change lead nitrate solution.

E.g. $$Na_2SO_3(s) + 2HCl(aq) \rightarrow 2NaCl(aq) + SO_2(g) + H_2O(l)$$
sodium sulphite + hydrochloric acid → sodium chloride + sulphur dioxide + water

Chloride(Cl^-)

When a chloride is treated with concentrated sulphuric acid, a colourless gas is produced. This gas (hydrogen chloride) forms steamy fumes in moist air and forms dense white fumes when mixed with ammonia gas.

$$NH_3(g) + HCl(g) \rightleftharpoons NH_4Cl(s)$$
ammonia + hydrogen chloride ⇌ ammonium chloride

If a chloride is mixed with manganese(IV) oxide and concentrated sulphuric acid added, the chloride and acid react as above to give HCl. This reacts with the manganese(IV) oxide, when the mixture is warmed, and a greenish-yellow gas is produced. This gas is chlorine and it turns blue litmus red and then bleaches it.

$$MnO_2(s) + 4HCl(aq) \rightarrow MnCl_2(aq) + 2H_2O(l) + Cl_2(g)$$
manganese(IV) oxide + hydrochloric acid → manganese(II) chloride + water + chlorine

When a solution of a chloride is acidified with dilute nitric acid and silver nitrate solution added, a white precipitate of silver chloride is formed immediately. This precipitate turns purple in sunlight and dissolves completely in concentrated ammonia solution.

E.g. $$NaCl(aq) + AgNO_3(aq) \rightarrow AgCl(s) + NaNO_3(aq)$$
sodium chloride + silver nitrate → silver chloride + sodium nitrate

Bromide (Br^-)

When a solution of a bromide is acidified with dilute nitric acid and silver nitrate solution added, a creamish precipitate of silver bromide is formed immediately. This precipitate dissolves partially in concentrated ammonia solution.

E.g. $$NaBr(aq) + AgNO_3(aq) \rightarrow AgBr(s) + NaNO_3(aq)$$
sodium bromide + silver nitrate → silver bromide + sodium nitrate

Iodide (I^-)
When a solution of an iodide is acidified with dilute nitric acid and silver nitrate solution added, a yellow precipitate of silver iodide is formed immediately. This precipitate is insoluble in ammonia solution.

E.g. $$NaI(aq) + AgNO_3(aq) \rightarrow AgI(s) + NaNO_3(aq)$$
sodium iodide + silver nitrate → silver iodide + sodium nitrate

Sulphate (SO_4^{2-})
When dilute hydrochloric acid and barium chloride solution are added to a solution of a sulphate, a white precipitate of barium sulphate is formed immediately.

E.g. $$Na_2SO_4(aq) + BaCl_2(aq) \rightarrow BaSO_4(s) + 2NaCl(aq)$$
sodium sulphate + barium chloride → barium sulphate + sodium chloride

Nitrate (NO_3^-)
There are two tests that can be used to test for a nitrate in solution.
(*i*) An equal volume of iron(II) sulphate solution (acidified with dilute sulphuric acid) is added to a suspected nitrate in a test tube. Concentrated sulphuric acid is poured carefully down the inside of the test tube, so that it forms a separate sulphuric acid layer below the aqueous layer. (This is because concentrated sulphuric acid is denser than water or aqueous solutions.) If a nitrate is present, a brown ring forms at the junction of the two layers. The brown substance is $FeSO_4.NO$ produced by the reduction of nitrate to nitrogen monoxide by iron(II) ions.

$$NO_3^-(aq) + 4H^+(aq) + 3Fe^{2+}(aq) \rightarrow NO(g) + 3Fe^{3+}(aq) + 2H_2O(l)$$
nitrate ions + hydrogen ions + iron(II) ions → nitrogen monoxide + iron(III) ions + water

This is called the **brown ring test**. Nitrites and bromides can give similar results.

(*ii*) Sodium hydroxide solution is added to a suspected nitrate and aluminium powder is added. (Sometimes **Devarda's alloy** is used in place of aluminium. This is an alloy containing aluminium that reacts more slowly than pure aluminium.) The mixture is warmed and hydrogen is produced. If a suspected nitrate is added, it will be reduced to ammonia gas. This will turn red litmus paper blue.

41.3 Tests for cations (positive ions)

There are three tests that can be used to identify cations.

(i) Flame tests
A small quantity of the compound is taken and a couple of drops of concentrated hydrochloric acid are added. A clean piece of platinum wire is dipped into the mixture and put into a hot Bunsen burner flame. Certain cations colour the Bunsen flame. Common flame colours are shown in Table 41.1.

Table 41.1 Flame tests for identifying cations

Flame colour	*Cation*
Orange-yellow	Sodium Na^+
Lilac-pink	Potassium K^+
Brick-red	Calcium Ca^{2+}
Pale-green	Barium Ba^{2+}
Green	Copper(II) Cu^{2+}
Blue	Lead Pb^{2+}

(ii) With sodium hydroxide solution
If a small quantity of the compound in solution is treated with sodium hydroxide solution,

an insoluble hydroxide may be precipitated. If a precipitate is formed, it may re-dissolve in excess sodium hydroxide solution. A summary of the precipitation of metal hydroxides with sodium hydroxide solution is shown in Table 41.2.

Table 41.2 Precipitation of metal hydroxides with sodium hydroxide

	Addition of sodium hydroxide solution	
Cation	*A couple of drops*	*Excess*
Potassium K^+	no precipitate	no precipitate
Sodium Na^+	no precipitate	no precipitate
Calcium Ca^{2+}	white precipitate	precipitate insoluble
Magnesium Mg^{2+}	white precipitate	precipitate insoluble
Aluminium Al^{3+}	white precipitate	precipitate soluble—colourless solution
Zinc Zn^{2+}	white precipitate	precipitate soluble—colourless solution
Iron(II) Fe^{2+}	green precipitate	precipitate insoluble
Iron(III) Fe^{3+}	red-brown precipitate	precipitate insoluble
Lead Pb^{2+}	white precipitate	precipitate soluble—colourless solution
Copper(II) Cu^{2+}	blue precipitate	precipitate insoluble
Silver Ag^+	grey-brown precipitate	precipitate insoluble

If no precipitate is formed, the solution is warmed. If the ammonium ion (NH_4^+) is present, ammonia gas is produced which turns red litmus blue.

E.g. $NH_4Cl(aq) + NaOH(aq) \rightarrow NH_3(g) + NaCl(aq) + H_2O(g)$
ammonium chloride + sodium hydroxide → ammonia + sodium chloride + water

(iii) With aqueous ammonia solution (ammonium hydroxide)

If a small quantity of the compound in solution is treated with ammonia solution, an insoluble hydroxide may be precipitated. If a precipitate is formed, it may re-dissolve in excess ammonia solution. A summary is shown in Table 41.3.

Table 41.3 Precipitation of metal hydroxides with ammonia solution

	Addition of ammonia solution	
Cation	*A couple of drops*	*Excess*
Potassium	no precipitate	no precipitate
Sodium	no precipitate	no precipitate
Calcium	no precipitate	no precipitate
Magnesium	white precipitate	precipitate insoluble
Aluminium	white precipitate	precipitate insoluble
Zinc	white precipitate	precipitate soluble—colourless solution
Iron(II)	green precipitate	precipitate insoluble
Iron(III)	red-brown precipitate	precipitate insoluble
Lead	white precipitate	precipitate insoluble
Copper	blue precipitate	precipitate soluble—blue solution
Silver	brown precipitate	precipitate soluble

41.4 Testing for gases

Table 41.4 summarises the tests for common gases.

Table 41.4 Summary of properties of common gases

Gas	*Formula*	*Colour*	*Smell*	*Test with moist litmus*	*Test with lighted splint*	*Other tests*
Hydrogen	H_2	×	×	×	Squeaky pop	
Oxygen	O_2	×	×	×	Relights glowing splint	
Nitrogen	N_2	×	×	×	Extinguished	Forms compound with magnesium
Chlorine	Cl_2	Greenish-yellow	✓	Blue → red then bleaches	Extinguished	
Hydrogen chloride	HCl	×	✓	Blue → red	Extinguished	White fumes with ammonia
Carbon dioxide	CO_2	×	×	Little change	Extinguished	Turns limewater milky
Carbon monoxide	CO	×	×	×	Burns with blue flame	
Ammonia	NH_3	×	✓	Red → blue	Extinguished	White fumes with hydrogen chloride
Sulphur dioxide	SO_2	×	✓	Blue → red	Extinguished	Turns potassium dichromate green. No effect on lead nitrate
Hydrogen sulphide	H_2S	×	✓	Blue → slightly red	Burns producing sulphur dioxide	Lead nitrate paper turns black
Nitrogen dioxide	NO_2	Brown	✓	Blue → red	Extinguished	
Dinitrogen monoxide	N_2O	×	✓	×	Relights glowing splint	Quite soluble
Nitrogen monoxide	NO	×	—	—	—	Forms brown fumes in air

42 The diaphragm cell

This process involves the electrolysis of a nearly saturated solution of sodium chloride using a carbon anode and a steel cathode (Fig. 42.1). The diaphragm is made of an asbestos substitute. It allows the solution to pass through but prevents the products, chlorine and sodium hydroxide solution, coming into contact.

The electrode reactions are:

$$\text{anode } 2Cl^- \rightarrow Cl_2 + 2e$$
$$\text{cathode } 2H^+ + 2e \rightarrow H_2$$

The solution dripping from the cathode consists of a solution of sodium hydroxide containing about 12% sodium chloride. The solution is concentrated by evaporation when most of the sodium chloride crystallises out.

The products are chlorine, hydrogen and sodium hydroxide. The sodium hydroxide produced is not as pure as that produced by the Kellner-Solvay process but the large initial expense of mercury is avoided.

When siting a factory for this process it would be advisable to have:

(*i*) a ready source of sodium chloride, *e.g.* salt deposits;
(*ii*) a good source of electricity;
(*iii*) associated industries to use sodium hydroxide and chlorine, *e.g.* soapmaking, bleaching cotton etc.

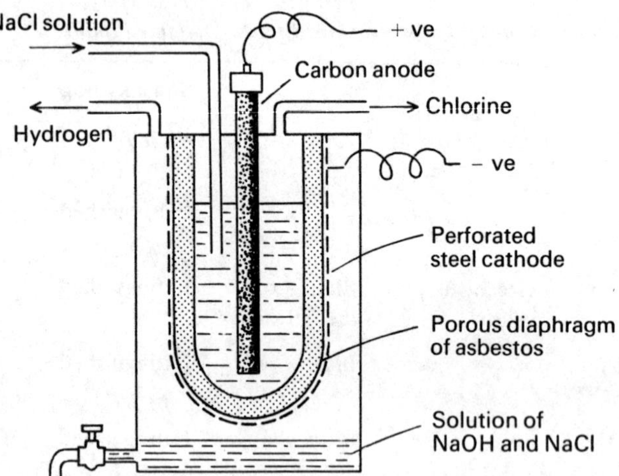

Fig. 42.1 The electrolysis of a nearly saturated solution of sodium chloride

Section III Self-test units

Do not attempt these tests until you have worked through the relevant unit in Section II. Answers appear on page 169.

Test on Unit 1 Elements, mixtures and compounds

1 A pure substance which cannot be split up by chemical reaction is called an ________.

Questions 2–4 refer to the following list: chlorine, sodium, helium, mercury, iodine. Select from the list:

2 A liquid element at room temperature and pressure.
3 A gaseous element at room temperature and pressure.
4 Two metallic elements.
5 The combination of elements together to form a compound is called ________.

Questions 6–7 refer to the following list: sodium oxide, sodium sulphide, sodium sulphate, sodium sulphite, sodium thiosulphate, sodium hydroxide. Select from the list:

6 Two compounds containing only two elements.
7 All compounds containing oxygen.
8 The percentage of calcium in a pure sample of calcium carbonate can vary from 30–50%. True or false?
9 Potassium cyanate has a formula KCNO. Name the four elements in this compound.
10 What is the formula of potassium thiocyanate?
11 Potassium nitrate has a formula KNO_3. What does the formula KNO_2 represent?
12 What compound does the formula Mg_3N_2 represent?

Name the elements present in:

13 Calcium hydroxide
14 Sodium hydrogensulphate.
15 Sodium hydride.

Test on Unit 2 Separation techniques in chemistry

Questions 1–5. Complete the following:

When salt is added to water, the salt ________ (**1**) to form a salt ________ (**2**). The impurities that do not dissolve are said to be ________ (**3**) in water and are removed by ________ (**4**). Salt crystals can then be obtained by ________ (**5**).

6 Pure water can be obtained from sea water by ________ (**6**).

Questions 7–15. Name the method most suitable for separating:

7 some solid sediment at the bottom of a bottle of wine;
8 unreacted copper(II) oxide from a solution of copper(II) sulphate solution;
9 water from a copper(II) sulphate solution;
10 propanone (boiling point 58°C) from a mixture of propanone and water;
11 ammonium chloride from a solid fertiliser;
12 oxygen from liquid air;
13 two amino acids in solution;
14 a mixture of paraffin and water;
15 solid pieces of metal from engine oil.

Test on Unit 3 Symbols, formulae and equations

(This is a very difficult but important unit and you would be advised to try this test more than once.)

Questions 1–7. Write down the names of the compounds represented by the following formulae:

1 $NaOH$
2 HNO_3
3 H_2SO_4
4 $Ca(HCO_3)_2$
5 $Mg(NO_3)_2$
6 $Zn(OH)_2$
7 $(NH_4)_2CO_3$

Questions 8–15. Write down the correct formulae for:

8 Silver chloride
9 Lead sulphide
10 Potassium carbonate
11 Iron(II) chloride
12 Iron(III) chloride
13 Iron(III) sulphate
14 Sodium sulphite (sodium Na^+, sulphite SO_3^{2-})
15 Barium phosphate (Barium Ba^{2+}, phosphate PO_4^{3-})

(If you can get 14 and 15 correct but not 8–13, it suggests that you have not remembered the list of ions in Table 3.1.)

Questions 16–20. Balance the following equations (State symbols omitted):

16 $Mg + HCl \rightarrow MgCl_2 + H_2$
17 $CuO + HCl \rightarrow CuCl_2 + H_2O$
18 $Ca + H_2O \rightarrow Ca(OH)_2 + H_2$
19 $Na + H_2O \rightarrow NaOH + H_2$
20. $ZnCl_2 + Na_2CO_3 \rightarrow ZnCO_3 + NaCl$

Test on Unit 4 Structure of the atom

Questions 1–5. Name the basic particle in the atom which:

1 has a negative charge;
2 has a positive charge;
3 is neutral;
4 has negligible mass;
5 is not found in the nucleus of an atom.
6 Atoms of the same element containing different numbers of neutrons are called ______.

Questions 7–10. Use the following information:

Lithium atomic no. = 3, mass no. = 7
Sodium atomic no. = 11, mass no. = 23
Potassium atomic no. = 19, mass no. = 39

7 How many protons, neutrons and electrons are present in a sodium atom?
8 How many protons, neutrons and electrons are present in a sodium Na^+ ion?
9 Write down the electron arrangements in lithium, sodium and potassium.
10 What similarity in electronic arrangement is seen with these elements?
11 An atom contains equal numbers of ______ and ______.
12 The smallest part of an element that can exist is called an ______.
13 All isotopes of calcium (atomic number 20) contain the same number of protons. True or false?
14 All isotopes are radioactive. True or false?
15 Why is the relative atomic mass of fluorine almost exactly 19?

Test on Unit 5 Bonding and structure

1 The existence of two or more forms of the same element in the same physical state is called ________.

2 A suitable refractory oxide is ________.

3 A refractory oxide has ________ bonding.

Questions 4–7. Are the following pairs allotropes?

4 Diamond and graphite. (Yes or no)

5 Carbon-12 and carbon-14. (Yes or no)

6 Liquid oxygen and oxygen gas. (Yes or no)

7 Oxygen gas O_2 and ozone gas O_3. (Yes or no)

8 The water molecule H_2O consists of two hydrogen atoms bonded to one oxygen atom. The bonds within the molecules are covalent bonds. True or false?

9 The bonds between the molecules are hydrogen bonds. True or false?

10 Covalent bonding involves ________ of electrons but ionic bonding involves a complete ________ of electrons.

Questions 11–15. Magnesium (electronic arrangement 2,8,2) and fluorine (2,7) combine to form a compound magnesium fluoride.

11 What electronic change takes place when a magnesium atom forms a magnesium ion?

12 What electronic change takes place when a fluorine atom forms a fluoride ion?

13 What type of bonding is present in magnesium fluoride?

14 Would magnesium fluoride be solid, liquid or gas at room temperature?

15 What is the formula of magnesium fluoride?

Test on Unit 6 States of matter

Questions 1–6. Fill in the spaces:

When a substance boils it changes state from ________ (1) to ________ (2). The temperature remains constant at the ________ ________ (3) during this change. When a substance freezes, it changes state from ________ (4) to ________ (5). The temperature at which this takes place is called the ________ (6).

Questions 7–9. Which state of matter:

7 is most easily compressed?

8 contains particles most closely packed together?

9 contains particles with the greatest average energy?

10 The spontaneous movement of particles to spread out to fill the whole container is called ________.

Test on Unit 7 Some laws relating to gases

Questions 1–6. Fill in the spaces:

The volume of a fixed mass of gas, at constant temperature, ________ (1) as pressure is increased. This is a statement of ________ (2) Law. The volume of a fixed mass of gas, at constant pressure, ________ (3) as the temperature increases. This is a statement of ________ (4) Law.

A gas whose volume is measured at stp is at a temperature of ________ (5) and a pressure of ________ (6).

7 Convert 23°C to an absolute (Kelvin) temperature.

Questions 8–9. A gas has a volume of 27.3 cm³ at 0°C and 1 atmosphere pressure. What would be the volume at:

8 0.5 atmospheres and 0°C?

9 1 atmosphere and 10°C?

10 $$\frac{P_1T_1}{V_1} = \frac{P_2T_2}{V_2}$$ True or false?

Test on Unit 8 Air and combustion

Questions 1–7. Name:

1 the gas present in air in largest amounts;
2 the gas which boils off first during the fractional distillation of liquid air;
3 two substances necessary before a piece of iron will rust;
4 the process in plants that converts carbon dioxide to oxygen;
5 two gases produced when a compound of carbon and hydrogen burns in excess air;
6 the chief chemical constituent of rust;
7 three ways of preventing iron from rusting.

Questions 8–12. 80 cm^3 of air were passed backwards and forwards over heated copper until no further change took place.

8 Name the gas removed when the air passed over the copper.
9 What volume should have remained at the end of the experiment?
10 The copper turns black during the experiment. What is the new chemical formed?
11 Why was the final volume of gas not measured until the apparatus had cooled to room temperature?
12 The increase in mass of the contents of the copper is equal to the loss of mass of the air. True or false?
13 Why is it dangerous to run a car engine in an enclosed garage?
14 ______ is the combination of a substance with oxygen.
15 The composition of air is fixed. True or false?

Test on Unit 9 Oxygen and oxides

Questions 1–5. Write the name and formula for:

1 the acid formed when carbon dioxide dissolves in water;
2 a neutral oxide;
3 a mixed oxide;
4 the acid anhydride of sulphurous acid;
5 the salt formed when zinc oxide reacts with sodium hydroxide solution.
6 Pure copper(II) oxide can be produced by heating a piece of copper in pure oxygen. True or false?
7 When a metal burns in air or oxygen it forms a ______ or ______ oxide.
8 When a non-metal burns in air or oxygen it may form an ______ oxide.
9 An ______ oxide can behave as either an acidic or basic oxide depending on conditions.
10 Oxygen can be prepared in the laboratory by the decomposition of ______ using a catalyst.

Test on Unit 10 Hydrogen

Questions 1–5. Name:

1 a metal which produces hydrogen with a dilute acid;
2 a metal which produces hydrogen with an alkali;
3 a suitable drying agent for hydrogen;
4 the substance produced when hydrogen burns in air;
5 a catalyst for the laboratory preparation of hydrogen.

Questions 6–10. These questions refer to the reduction of copper(II) oxide with hydrogen (Fig. 13.1).

6 Why must care be taken when lighting the excess hydrogen escaping through the jet?
7 What is the solid product formed in the test tube?
8 During the early stages of the experiment, droplets of a colourless liquid may be seen in the test tube near the jet. What is this liquid?
9 Why must the apparatus be allowed to cool down, with the hydrogen passing through, before being dismantled?
10 Name two other metal oxides that can be reduced in a similar way.

Test on Unit 11 Water

Questions 1–5. Select from the following list: calcium hydroxide, calcium carbonate, calcium stearate, calcium sulphate, calcium hydrogencarbonate, anhydrous calcium chloride. Name the substance which:

1 is formed as scum when soap is added to hard water;
2 may be present in permanent hard water;
3 is formed when calcium reacts with cold water;
4 is present in temporary hard water;
5 is deliquescent.
6 An efflorescent compound contains water of crystallisation. True or false?
7 A sample of a washing powder lathers well in hard water without forming scum. Is it a soap or a soapless detergent?

Question 8 refers to Fig. 11.1.

8 Which of the substances is most soluble in water at
(*i*) 20°C and (*ii*) 30°C?
9 If anhydrous copper(II) sulphate is added to a colourless liquid and it turns blue this proves that the liquid is pure water. True or false?
10 Chlorine is added to water from reservoirs in order to remove the hardness. True or false?

Test on Unit 12 The mole and chemical calculations

Relative atomic masses used in this test: $H = 1$, $C = 12$, $O = 16$, $Mg = 24$, $S = 32$, $Cl = 35.5$, $K = 39$, $Cu = 64$. (*1 mole of molecules of any gas occupies 24 000* cm^3 *at room temperature and pressure.*)

1 1 mole of hydrogen molecules H_2 contains 6×10^{23} atoms (Avogadro's number) of hydrogen. True or false?

Questions 2–4. A sample of an oxide of copper (mass 7.2 g) was found to contain 6.4 g of copper.

2 What mass of oxygen was present in the sample?
3 What mass of oxygen would combine with 64 g of copper in this oxide?
4 What is the simplest formula for this oxide?

Questions 5–7. The formula of potassium hydrogencarbonate is $KHCO_3$.

5 What is the mass of 1 mole of potassium hydrogencarbonate?
6 What is the percentage of oxygen in this compound?
7 What mass of potassium hydrogencarbonate would be required to prepare 100 cm^3 of 0.1 M solution?
8 How many times heavier is 1 atom of magnesium than 1 atom of carbon?
9 How many moles of sulphur dioxide molecules SO_2 are present in 16 g of sulphur dioxide?
10 An oxide of sulphur contains 40% sulphur and 60% oxygen. What is the simplest formula for the oxide?

Questions 11–12.

$$CuSO_4.5H_2O(s) \rightarrow CuSO_4(s) + 5H_2O(g)$$

11 Calculate the mass of 1 mole of copper sulphate crystals $CuSO_4.5H_2O$.
12 Calculate the mass of anhydrous copper sulphate $CuSO_4$ produced by heating 50 g of copper sulphate crystals.

Questions 13–15.

$$Mg(s) + 2HCl(aq) \rightarrow MgCl_2(aq) + H_2(g)$$

13 What volume of hydrogen gas, at room temperature and pressure, is produced when 1 mole of magnesium atoms has reacted with excess hydrochloric acid?
14 Calculate the volume of hydrogen, at room temperature and pressure, produced by 1 g of magnesium reacting with excess hydrochloric acid.
15 Calculate the volume of M hydrochloric acid required to react with 1 g of magnesium.

Test on Unit 13 Laws of chemical combination

Part A Laws relating to reacting masses

1 $Na_2CO_3(s) + 2HCl(aq) \rightarrow 2NaCl(aq) + H_2O(l) + CO_2(g)$
If the above reaction was carried out in an open beaker, the mass of the beaker and contents would remain unchanged throughout. True or false?

Questions 2–3. Two oxides of chromium were analysed. Oxide X. 10.4 g of chromium combined with 4.8 g of oxygen. Oxide Y. 2.6 g of chromium combined with 2.4 g of oxygen.

2 What is the ratio of chromium combining with a fixed mass of oxygen in oxides *X* and *Y*?
3 Which chemical law does this verify?

Questions 4–5. There are different methods available for preparing a pure sample of iron(III) oxide.

4 The percentage of iron in the different samples produced by different methods, is the same. True or false?
5 This verifies the law of conservation of mass. True or false?

Part B Laws relating to combining volumes of gases

(In all questions, the volumes of gases are measured at room temperature and pressure.)

6
$$CO_2(g) + C(s) \rightarrow 2CO(g)$$
What is the volume of carbon monoxide produced when 40 cm^3 of carbon dioxide is passed over heated carbon?

Questions 7–10 concern the decomposition of ammonia into a mixture of nitrogen and hydrogen and the analysis of the resulting mixture.

$$2NH_3(g) \rightarrow N_2(g) + 3H_2(g)$$

40 cm^3 of ammonia gas were passed over heated iron to produce a mixture of nitrogen and hydrogen.

7 What was the volume of nitrogen produced?
8 What was the volume of hydrogen produced?
9 What was the volume of the mixture of nitrogen and hydrogen?
10 The resulting mixture was passed over heated copper(II) oxide

$$CuO(s) + H_2(g) \rightarrow Cu(s) + H_2O(g)$$

What is the final volume of gas after passing over heated copper(II) oxide?

Test on Unit 14 Metals(I)

Questions 1–5. Complete the following passage:

Sodium hydroxide is produced by the electrolysis of __________ (1) using a __________ (2) cathode. The sodium, discharged at the cathode, produces a sodium __________ (3), which reacts with water to produce sodium hydroxide. By-products of this process are the gases __________ (4) and __________ (5).

Questions 6–10. Select the metal most suitable from the following list:
sodium, calcium, aluminium, copper, lead, mercury.

6 A metal to make plates for a car battery.
7 A metal for filling a thermometer.
8 A metal for making a fishing weight.
9 A metal for making lightweight electricity cables that will require the minimum number of pylons for support.
10 A metal for mixing with tin to make solder.

Questions 11–14. Complete the following passage:

Copper can be purified by electrolysis using copper(II) sulphate solution. A pure copper plate is used for the __________ (11) electrode and the impure copper plate is used for the __________ (12) electrode. Pure copper is deposited on the __________ (13) electrode. The mass of the negative electrode __________ (14) during the electrolysis.

15 A mixture of metals is called an __________.

Test on Unit 15 Metals(II)

1 When the electrode potential of different metals were separately compared with a standard copper plate the following results were obtained.

Metal plate	Aluminium	Magnesium	Lead	Silver	Iron
Potential difference (volts)	−2.00	−2.71	−0.47	+0.46	−0.78

Complete the list below by placing all the above metals in their correct order in the electrochemical series.

(*a*) ________
(*b*) ________
(*c*) ________
(*d*) ________
(*e*) Copper
(*f*) ________

(*East Anglian Examinations Board—South*)

Questions 2–7. The following list of metals should be used to answer these questions:

Potassium (most reactive)
Calcium
Zinc
Copper
Mercury

Name:

2 a metal which could be extracted from its ore by heating alone;
3 two metals that are usually extracted from their ores by electrolysis;
4 the metal which forms the most stable compounds;
5 two metals that do not react with dilute hydrochloric acid.

Questions 6–7. Excess zinc powder was put into a blue solution of copper(II) sulphate solution. After some time the reaction was complete.

6 What are the products of the reaction?
7 What would be seen during the reaction?

Questions 8–11. Complete the following passage:

Sodium is extracted from ________ **(8)** sodium chloride by electrolysis. Calcium chloride is added to ________ **(9)** the melting point. Sodium is discharged at the ________ **(10)** electrode and chlorine at the ________ **(11)** electrode.

Questions 12–15. Complete the following passage:

Iron is extracted from iron ore in a blast furnace and the furnace is heated by ________ **(12)**. Iron ore, ________ **(13)** and ________ **(14)** are loaded into the furnace continuously. The iron oxide is reduced to iron by ________ **(15)**.

Test on Unit 16 Oxidation and reduction

Questions 1–6. Which substance is oxidised and which substance is reduced in each of the following equations?

1 $2Al(s) + Fe_2O_3(s) \rightarrow Al_2O_3(s) + 2Fe(s)$
aluminium + iron(III) oxide → aluminium oxide + iron

2 $Cl_2(g) + 2I^-(aq) \rightarrow I_2(aq) + 2Cl^-(aq)$
chlorine + iodide ion → iodine + chloride ion

3 $Mg(s) + 2H^+(aq) \rightarrow Mg^{2+}(aq) + H_2(g)$
magnesium + hydrogen ion → magnesium ion + hydrogen

4 $Mg(s) + CuO(s) \rightarrow MgO(s) + Cu(s)$
magnesium + copper(II) oxide → magnesium oxide + copper

5 $Cu^{2+}(aq) + Zn(s) \rightarrow Cu(s) + Zn^{2+}(aq)$
copper(II) ion + zinc → copper + zinc ion

6 $CO_2(g) + C(s) \rightarrow 2CO(g)$
carbon dioxide + carbon → carbon monoxide

7 Name the reducing agent in question **1**.
8 Name the reducing agent in question **2**.
9 If a substance loses hydrogen during a reaction, it is being reduced. True or false?
10 Substances containing oxygen are always oxidising agents. True or false?

Test on Unit 17 The effect of electricity on chemicals
Part A Qualitative electrolysis

Questions 1–11. Complete the following:

Solid potassium iodide (KI) does not conduct electricity because the ions are held together in a rigid crystal ________ **(1)** and are not free to move. When the solid is melted a current passes because ________________ **(2)** are produced. This passage of electricity decomposes the molten potassium iodide into potassium and iodine. Potassium iodide is called the ________ **(3)** and the decomposition using electricity is called ________ **(4)**.

During this passage of electricity through molten potassium iodide, ________ **(5)** is produced at the positive electrode (or ________ **(6)**) and ________ **(7)** is produced at the negative electrode (or ________ **(8)**).

Electrolysis of an aqueous solution of potassium iodide produces ________ **(9)** at the positive electrode and ________ **(10)** at the negative electrode. An aqueous solution of potassium iodide contains the following ions: ________ **(11)**.

Part B Quantitative electrolysis

Questions 12–15. A current of 5 A passed through molten calcium bromide ($CaBr_2$) for 3 minutes 13 seconds.

12 Calculate the number of coulombs of electricity that have passed.
13 Calculate the number of Faradays that have passed. (1 Faraday = 96 500 coulombs.)
14 Calculate the mass of calcium deposited at the negative electrode. (Calcium ion Ca^{2+}, Ca = 40.)
15 Calculate the mass of bromine liberated at the positive electrode. (Bromide ion Br^-, Br = 80.)

Test on Unit 18 Chemical families and the Periodic Table

Part A Chemical families

Questions 1–4. The alkali metal family includes the elements lithium, sodium, potassium, rubidium and caesium.

1 The least reactive of these five metals is ______________________________ .

2 All alkali metals are stored in ____________ to exclude air and water.

3 When a piece of lithium is burnt in oxygen, the product is ______________________ .

4 When a piece of lithium is added to cold water, the products are ____________ and ________.

Questions 5–7. The halogen family includes fluorine, chlorine, bromine and iodine.

5 Name a solid halogen at room temperature and pressure.

6 Name the solid formed when bromine vapour is passed over heated iron.

7 In which one of the following does a reaction *not* take place?

(*i*) potassium bromide + iodine
(*ii*) potassium iodide + bromine
(*iii*) potassium iodide + chlorine

Part B The Periodic Table

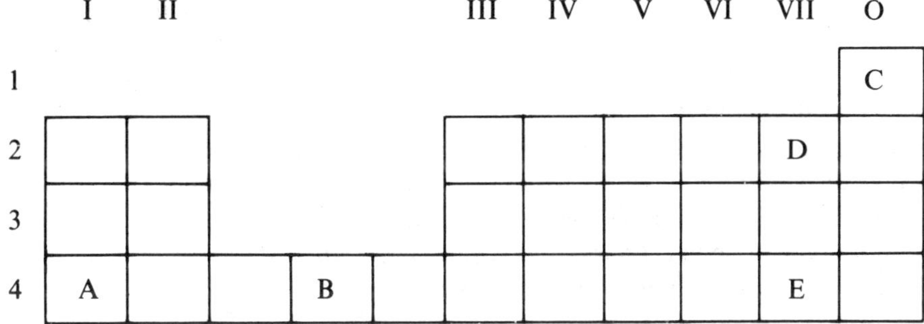

(The letters A–E are not symbols for elements.)

Questions 8–15. The skeleton Periodic Table above should be used to answer the following questions.

Write down the letters for:

8 two elements in the same group;

9 an alkali metal;

10 a noble gas;

11 a transition metal;

12 the element containing the greatest number of protons;

13 the element containing the smallest atoms;

14 the element whose atoms contain only one electron in the outer energy level.

15 What type of bonding would be expected in a compound between A and D?

Test on Unit 19 Acids, bases and salts

Questions 1–3. Choose your answers from the following: calcium hydroxide, barium chloride, sodium carbonate, sodium hydrogencarbonate, silver chloride, zinc carbonate.
From the list above select:

1 an acid salt;
2 an alkali;
3 two insoluble salts.

Questions 4–7. Complete the following word equations:

4 Zinc oxide + ________ → zinc sulphate + ________
5 ________ + nitric acid → calcium nitrate + ________ + ________
6 ________ + ________ → zinc chloride + hydrogen
7 Barium nitrate + sodium sulphate → ________ + ________
8 Which one of the reactions in questions 4–7 is an example of double decomposition?
9 A base can be defined as a proton acceptor. True or false?
10 Name a soluble lead salt.

Questions 11–15. Copper(II) sulphate crystals, $CuSO_4.5H_2O$, can be produced using copper(II) oxide and dilute sulphuric acid.

Excess copper(II) oxide was added in small amounts to warm dilute sulphuric acid. The solution was filtered into an evaporating basin. The solution was evaporated carefully until crystals formed on a glass rod which had been dipped into the solution. The solution was then allowed to cool and crystallise.

11 Why was the acid heated before the solid was added?
12 Why was excess copper(II) oxide used?
13 Why was the solution filtered before evaporation?
14 Why is it not advisable to evaporate the solution to dryness?
15 Write an equation in symbols for the reaction.

Test on Unit 20 Rates of reaction

The following results were obtained when 0.24 g of magnesium ribbon were added to an excess of dilute hydrochloric acid while the temperature was kept constant.

Time (seconds)	Volume of gas evolved (cm^3)
0	0
20	90
40	140
60	172
80	195
100	210
120	224
140	224

1 Name the gas evolved and write the chemical equation for the reaction.
2 What was the volume of gas evolved in
(*i*) the first 20 second interval (from 0 seconds to 20 seconds),
(*ii*) the second 20 second interval (from 20 to 40 seconds),
(*iii*) the third 20 second interval (from 40 to 60 seconds)?
3 Explain why the volume of gas changes in each of these 20 second intervals.
4 Why is the volume of gas the same after 140 seconds and 120 seconds?
5 How would the initial rate of evolution of the gas change if
(*i*) the temperature was increased,
(*ii*) the volume of acid were diluted with an equal volume of water,
(*iii*) the same mass of magnesium powder were used instead of magnesium ribbon?
6 How would the final volume of gas change if
(*i*) the temperature of the gas were increased and the pressure kept constant,
(*ii*) the pressure on the gas were increased and the temperature kept constant,

(*iii*) a greater mass of magnesium were used with the same volume of acid,
(*iv*) a larger volume of acid were used with the same mass of magnesium?
(*East Anglian Examinations Board—South*)

Test on Unit 21 Reversible reactions and equilibrium

Questions 1–3 refer to the following reaction:

$$BiCl_3(aq) + H_2O(l) \rightleftharpoons BiOCl(s) + 2HCl(aq)$$

bismuth chloride + water ⇌ bismuth oxychloride + hydrochloric acid

A solution of bismuth chloride was prepared by adding solid bismuth chloride to concentrated hydrochloric acid.

1 What would be observed if water was added to this solution of bismuth chloride?
2 In which direction (left or right) does the equilibrium move when water is added?
3 Suggest one method of reversing this equilibrium change.
4 An equilibrium can only be established in reversible reactions. True or false?

Questions 5–8 refer to the following reaction:

$$H_2(g) + I_2(g) \rightleftharpoons 2HI(g)$$

hydrogen + iodine ⇌ hydrogen iodide

The forward reaction is endothermic. A mixture of hydrogen, iodine and hydrogen iodide were in equilibrium. What would be the effect on the equilibrium (move to left, move to right or unchanged) of each of the following changes?

5 Adding additional hydrogen to the equilibrium mixture.
6 Removing hydrogen iodide from the equilibrium mixture.
7 Increasing the pressure.
8 Increasing the temperature.

Questions 9–10 refer to the reaction:

$$N_2(g) + 3H_2(g) \rightleftharpoons 2NH_3(g)$$

The table below shows the effects of changes of temperature and pressure on the yield of ammonia.

	Percentage of ammonia in equilibrium mixture at pressures of:			
Temp. °C	1 atmosphere	100 atmospheres	200 atmospheres	1000 atmospheres
200	15.3	80.6	85.8	98.3
400	0.44	25.1	36.3	79.8
600	0.05	4.5	8.3	31.4
800	0.01	1.2	2.2	—
1000	0.004	0.4	0.9	

Using the table:

9 What is the effect of increasing pressure on the equilibrium?
10 What is the effect of increasing temperature on the equilibrium?

Test on Unit 22 Sulphur

Questions 1–4. There are various forms of sulphur. These include: α-sulphur, β-sulphur, plastic sulphur. Which one of these forms of sulphur:

1 does not consist of an arrangement of S_8 rings?
2 is most stable at room temperature?
3 is formed when a solution of sulphur evaporates at 40°C?
4 is formed when molten sulphur is rapidly cooled by pouring into cold water?
5 Name the compound formed when a mixture of iron and sulphur is heated.

Questions 6–10. Complete the following passage:

Sulphur is extracted from underground deposits by the ________ **(6)** process. A hole is drilled down to the deposits and a pump is placed in the hole. Superheated ________ **(7)** at 170°C is pumped down the ________ **(8)** tube to ________ **(9)** the sulphur. Hot compressed ________ **(10)** is pumped down the middle pipe to force the sulphur to the surface.

Test on Unit 23 Sulphides and hydrogen sulphide

Questions 1–3. When hydrogen sulphide was passed through a solution of zinc sulphate, a white precipitate was formed.

1 Name the white precipitate.
2 Write a symbol equation for this reaction.

The precipitate was filtered off and added to dilute sulphuric acid.

3 Name the gas evolved.
4 The salts produced by replacing hydrogens in hydrogen sulphide are called ________.
5 Hydrogen sulphide cannot be dried using concentrated sulphuric acid. True or false?
6 Natural gas has to have hydrogen sulphide removed before it can be used for combustion. Name the product of combustion of hydrogen sulphide which can lead to pollution problems.

Questions 7–10. Name the products of the following reactions:

7 Hydrogen sulphide and iron(III) chloride.
8 Hydrogen sulphide and chlorine.
9 Hydrogen sulphide and lead(II) nitrate.
10 Hydrogen sulphide and sulphur dioxide.

Test on Unit 24 Oxides of sulphur

Questions 1–7. Identify the substances A–G in the following statements:

1 A colourless gas A, with a pungent odour, when passed through an orange solution of potassium dichromate, turns the solution green, and forms a white precipitate.
2–3 A colourless gas B forms needle-shaped crystals when cooled in an ice-salt mixture. When these crystals are added to water, a strongly acidic solution C is produced.
4 When the mixture, resulting from the preparation of sulphur dioxide, is poured into water a blue solution D is produced.
5 A yellowish solid E is formed when burning magnesium is added to sulphur dioxide.
6 A white precipitate F is formed when sulphur dioxide is passed through limewater (calcium hydroxide).
7 A colourless gas G turns potassium dichromate green and does not form any precipitate.
8 Sulphur dioxide can be prepared by heating copper and dilute sulphuric acid. True or false?
9 Sulphur dioxide can act as an oxidising agent or a reducing agent. True or false?
10 Sulphur trioxide can be produced by burning sulphur in excess oxygen. True or false?

Test on Unit 25 Sulphuric acid and sulphates

Questions 1–5. Identify the substances A–E in the following statements:

1 A black solid residue A formed when concentrated sulphuric acid and sugar are mixed.

2 A colourless gas B produced when concentrated sulphuric acid is added to solid calcium chloride.

3 A pungent gas C produced when concentrated sulphuric acid and sulphur are heated together.

4 A red solid D formed when iron(II) sulphate crystals are strongly heated.

5 A solution E turns blue litmus red and gives a white precipitate with barium chloride solution.

Questions 6–10 concern the industrial manufacture of sulphuric acid.

6 Name three sources of sulphur dioxide in the process.

7 Why is the sulphur dioxide purified before being used in this process?

8 Write the symbol equation for the step converting sulphur dioxide into sulphur trioxide.

9 What is the catalyst for this step?

10 Sulphur dioxide is dissolved in concentrated sulphuric acid to form ________. ________ is added to this to produce concentrated sulphuric acid.

Test on Unit 26 Nitrogen

Questions 1–7. Complete the following passage:

In the preparation of nitrogen from the air, air is passed through ________ **(1)** solution to remove carbon dioxide and over heated ________ **(2)** to remove oxygen. The resulting gas is impure nitrogen and the chief impurities are the ________ **(3)** gases.

Pure nitrogen can be prepared by heating a mixture of ________ **(4)** and ________ **(5)**. This produces ________ **(6)** *in situ*. The gas is usually collected over ________ **(7)**.

Questions 8–9. What is necessary to bring about the following changes?

8 Nitrogen and oxygen → nitrogen monoxide.

9 Ammonia in the soil → nitrates in the soil.

10 Write the symbol equation for the burning of magnesium in nitrogen.

Test on Unit 27 Ammonia and ammonium compounds

Questions 1–5. Identify the substances A–E in the following statements:

1 A colourless solution A turns red litmus blue. On evaporation, no solid residue remains.

2 A colourless liquid B formed after cooling the products of the reaction between ammonia and copper(II) oxide.

3 A white solid C formed when ammonia and hydrogen chloride gases mix.

4 A colourless gas D evolved when a mixture of ammonium sulphate and sodium hydroxide is heated.

5 A red-brown precipitate E formed when ammonia solution is added to iron(III) chloride solution.

6 All ammonium compounds contain the NH_3^+ ion. True or false?

7 All ammonium compounds are soluble in water. True or false?

Questions 8–10. Complete the following passage.

Ammonia is produced by the ________ **(8)** process. A mixture of nitrogen and ________ **(9)** is passed over a heated catalyst of ________ **(10)** at 500°C.

Test on Unit 28 Nitric acid and nitrates

1 Nitric acid is prepared in the laboratory by heating potassium nitrate with ________.

2 Thermal decomposition of the nitric acid during the preparation causes it to be ________ in colour.

3 This colour can be removed by passing ________ through the acid.

Questions 4–8. Identify the substances A–E in the following statements:

4 A metal A which produces hydrogen with cold, dilute nitric acid.

5 A white crystalline solid B which decomposes on heating to produce a residue which is yellow when hot and white when cold. A brown gas is evolved which turns blue litmus red.

6 A blue solution is produced when a black solid C is added to warm nitric acid.

7 A colourless oily liquid D decomposes on strong heating to produce oxygen. When copper is added to this liquid brown fumes are produced.

8 When white crystals are heated, oxygen is evolved. The liquid remaining cools to produce a white solid E.

9 Concentrated nitric acid will dissolve silver. True or false?

10 When nitric acid acts as an oxidising agent, oxides of nitrogen are usually produced. True or false?

Test on Unit 29 Oxides of nitrogen

Questions 1–10. Name the oxide of nitrogen which:

1 relights a glowing splint;

2 contains the greatest percentage of nitrogen;

3 forms brown fumes in contact with air;

4 dissolves in water to form an acidic solution;

5 is prepared by heating ammonium nitrate;

6 forms a pale yellow liquid on cooling;

7 is brown in colour;

8 is prepared by heating lead(II) nitrate;

9 is least soluble in water;

10 forms a brown complex with iron(II) sulphate solution.

Test on Unit 30 Agricultural chemistry

1 Name three elements required in large amounts by a growing plant.

2 Iron, boron and copper are required in small amounts. They are called ________ elements.

3 Name a natural source of nitrogen fertiliser.

4 Very ________ fertilisers are quickly washed out of the soil.

5 Why is calcium phosphate, without treatment, not a good phosphorus fertiliser?

6 Lime and an ammonium fertiliser should not be used together because they would react together to produce ________.

7 Nitrogenous fertilisers are absorbed through the roots in the form of ________.

Questions 8–10. Which vital element is supplied to the soil by:

8 urea?

9 wood-ash?

10 bone meal?

Test on Unit 31 Phosphorus

1 Name three substances added to the furnace for the extraction of phosphorus.
2 Phosphorus is extracted by electrolysis using carbon electrodes. True or false?

Questions 3–5. When phosphorus burns in excess oxygen, a white solid is produced.

3 Name the white solid produced.
4 Name the substance formed when this white solid is dissolved in water.
5 Is the solution of the white solid in water acidic, alkaline or neutral?
6 Both allotropes of phosphorus are composed of P_4 tetrahedra. True or false?
7 How is white phosphorus stored?
8 Which form of phosphorus is formed when red phosphorus sublimes in an inert atmosphere at 431°C?
9 Phosphorus vapour condenses under water to form ________ phosphorus.
10 Salts containing phosphorus and oxygen are called ________________

Test on Unit 32 Hydrogen chloride and chlorine

Questions 1–4. Complete the following passage:

Hydrogen chloride gas can be prepared by the action of ________ (1) sulphuric acid on ________ (2). If manganese(IV) oxide is added to the mixture, and the mixture heated, ________ (3) gas is produced. Manganese(IV) oxide ________ (4) the hydrogen chloride.

Questions 5–9. Name the products of the reactions between:

5 chlorine and cold, dilute sodium hydroxide solution;
6 turpentine and chlorine;
7 iron and hydrogen chloride;
8 manganese(IV) oxide and concentrated hydrochloric acid;
9 hydrogen sulphide and chlorine.
10 When chlorine water is exposed to sunlight, oxygen is produced from the decomposition of chloric(I) acid (HOCl). Write a symbol equation for this decomposition.

Test on Unit 33 Carbon, carbon monoxide and organic chemistry

Part A Carbon

1 Draw diagrams to show the arrangement of carbon atoms in diamond and graphite.
2 Burning graphite in excess oxygen produces the same product as burning diamond in excess oxygen. True or false?

Questions 3–4 concern the reaction between copper(II) oxide and carbon.

3 Write a symbol equation for the reaction.
4 What is the function of the carbon?
5 All bonds between carbon atoms in a diamond are covalent. True or false?

Part B Carbon monoxide

1 Carbon monoxide is produced during incomplete combustion of carbon compounds. True or false?

Questions 2–3 concern the reaction taking place when carbon dioxide is passed over heated carbon. Some of the carbon dioxide is converted to carbon monoxide.

2 Write a symbol equation for this reaction.
3 How could excess carbon dioxide be removed from the mixture of carbon dioxide and carbon monoxide?
4 Write a symbol equation for the reaction between iron(III) oxide and carbon monoxide.
5 Name the gas produced when carbon monoxide burns in air.

Part C Organic chemistry

Questions 1–3. Complete the following passage:

Crude oil can be split up into fractions with different boiling points by __________ __________ (1). The higher boiling point fractions can be split up into molecules with shorter chain lengths by __________ (2). When liquid paraffin vapour is passed over heated broken china, the gas __________ (3) is produced.

Questions 4–6. Name a reagent which will detect:

4 starch;
5 a reducing sugar;
6 an unsaturated compound.

Questions 7–10. Five organic compounds (labelled A–E) are shown below. They should be used to answer the following questions.

```
     H              H  H  H              H       H  H
     |              |  |  |              |       |  |
   H—C—H         H—C—C—C—OH          H—C—O—C—C—H
     |              |  |  |              |       |  |
     H              H  H  H              H       H  H

     A                 B                        C

            H  H             H       H
            |  |              \     /
          H—C—C—H              C=C
            |  |              /     \
            H  H             H       H

              D                  E
```

Select:

7 one unsaturated compound;
8 two isomers of the same molecular formula;
9 two homologues of the same homologous series;
10 one compound which forms an alkene on dehydration.

Test on Unit 34 Carbon dioxide and carbonates

1 Name two natural forms of calcium carbonate.

Questions 2–3. When calcium carbonate is strongly heated its mass decreases and a white solid residue remains. When this is added to water, an alkaline solution remains.

2 What is the chief chemical constituent of the white solid residue?
3 What is the alkaline solution produced?

Questions 4–8. Identify the substances A–E.

4 When carbon dioxide is bubbled through a saturated solution of sodium carbonate, a white precipitate A is formed.
5 When a bluish-green solid B is heated, it turns black and evolves a colourless gas which turns limewater milky and is odourless.
6 A sodium compound C evolves carbon dioxide when heated.
7 When carbon dioxide is passed through calcium hydroxide solution, a white precipitate D is formed.
8 When burning magnesium ribbon is put into carbon dioxide, black specks of E are formed.

Questions 9–10.

H_2CO_3	carbonic acid	pH 5.5
Na_2CO_3	sodium carbonate	pH 10.5

9 Write down the name and formula for the acid salt of carbonic acid containing sodium.
10 What is the approximate pH of a solution of the acid salt?

Test on Unit 35 Polymers

1 A reaction in which two molecules combine together with the resulting elimination of a small molecule is called a ________ reaction.
2 A reaction in which two molecules combine to form a single product is called an ________ reaction.
3 Name a thermosetting polymer.
4 Name a thermoplastic polymer.
5 All monomers for addition polymers contain a ________ bond between two carbon atoms.
6 The monomers for condensation polymerisation must contain two reactive groups. True or false?
7 An addition polymer has the same empirical formula as the monomer from which it is composed. True or false?
8 Name the monomer used in the production of polythene.
9 The properties of a polymer are independent of the conditions of polymerisation. True or false?
10 Hexane-1,6-diamine and hexanedioic acid are monomers for producing nylon-6,6. True or false?

Test on Unit 36 Fuels

1 Coal is a pure form of carbon. True or false?
2 What is the chief product formed when coal burns in excess air?
3 Name four substances produced when coal is heated in the absence of air.
4 What name is given to the process in **3**?
5 How was the coal formed in the earth?
6 What is the chief chemical constituent of natural gas?
7 What is the name of the gaseous fuel produced when air is passed over heated carbon? What are the chief chemical constituents of this gas?
8 What is the name of the gaseous fuel produced when steam is passed over heated carbon? What are the chief chemical constituents of this gas?
9 Why is the gaseous fuel in **7** not as good as the gaseous fuel in **8**?
10 In the space shuttle, two liquid fuel rockets are used to boost the space shuttle into orbit. Two suitable fuels are shown below.

	Hydrogen H_2	Hydrazine N_2H_4
Heat of combustion	−286 kJ/mole	−90 kJ/mole
Mass of 1 mole	2 g	32 g

Which one of these two fuels would you recommend on the basis of the information given? Give a reason for your answer.

Test on Unit 37 Energy in chemistry

Questions 1–3. The heat produced from the combustion of 8 g of sulphur raised the temperature of 100 g of water by 10°C.

1 Calculate the amount of heat produced by the combustion of 8 g of sulphur. (Specific heat capacity of water = 4.2 kJ/mole/deg.)
2 Calculate the molar heat of combustion of sulphur (S = 32).
3 What is the most likely reason for the discrepancy between this result and the value in the data book?
4 During an exothermic reaction, the products contain more energy than the reactants. True or false?
5 A fuel cell is a means of converting ________ energy into ________ energy.
6 The heat of combustion of carbon is −393.5 kJ. Calculate the heat evolved when 60 g of carbon is converted to carbon dioxide. (C = 12.)
7 The heat liberated or absorbed when 1 mole of a substance is formed from its constituent elements is called the heat of ____________.

8 Iodine and chlorine react as follows:

$$I_2(s) + Cl_2(g) \rightarrow 2ICl(s) \quad (\Delta H = -68 \text{ kJ})$$

The heat of formation of iodine monochloride is ______ kJ/mole.

Questions 9–10. The equation for the fermentation of glucose is:

$$C_6H_{12}O_6(s) \rightarrow 2C_2H_5OH(l) + 2CO_2(g)$$

The heats of formation of glucose, ethanol and carbon dioxide are −1260 kJ, −278 kJ and −395 kJ respectively.

9 Is the reaction above exothermic or endothermic?
10 Calculate the heat of reaction for the above reaction.

Test on Unit 38 Social and economic considerations

In this test cross-references are made to other Units in Section II.

1 Reference Unit 10.4. Why is method (*ii*) more frequently used today than method (*i*) in the United Kingdom?
2 Reference Unit 36.2. What particular pollutant is produced by burning coal?
3 Reference Unit 33.21. Why is cracking of hydrocarbons an important economic process?
4 Reference Unit 34.6. Name two materials that can be recycled in the Solvay process.
5 Give an example of waste gases from a furnace being used to heat the furnace.

Test on Unit 39 Radioactivity

Questions 1–5. Which type of radiation:

1 is not deflected by magnetic fields?
2 consists of a stream of electrons?
3 has the lowest penetrating power?
4 can be readily detected using laboratory apparatus?
5 consists of a stream of helium nuclei?

Questions 6–8. A radioactive isotope has a half life of 2 hours. The background radiation was 12 counts per minute. The initial count was 140 counts per minute.

6 What was the initial count after correction for background radiation?
7 What would be the corrected reading after 4 hours?
8 What would be the observed reading after 4 hours?
9 The isotope $^{227}_{89}Ac$ decays by β decay. What is the atomic number and mass number of the isotope produced?
10 The isotope $^{214}_{84}Po$ decays by α decay. What is the atomic number and mass number of the isotope produced?

Test on Unit 40 Quantitative volumetric chemistry

Questions 1–4.

$$HCl(aq) + NaOH(aq) \rightarrow NaCl(aq) + H_2O(l)$$

25 cm^3 of 0.1 M *sodium hydroxide solution is titrated with 0.08* M *hydrochloric acid.*

1 Calculate the mass of sodium hydroxide required to make 1 dm^3 of 0.1 M sodium hydroxide solution.

2 Calculate the number of moles of sodium hydroxide present in 25 cm^3 of 0.1 M sodium hydroxide solution.

3 How many moles of hydrochloric acid react with 25 cm^3 of 0.1 M sodium hydroxide solution?

4 Calculate the volume of 0.08 M hydrochloric acid reacting with 25 cm^3 of 0.1 M sodium hydroxide solution.

5 During a titration water can be added to the solution in the burette without affecting the results. True or false?

Test on Unit 41 Qualitative chemistry

Questions 1–5. Name:

1 a coloured gaseous compound;

2 a gas that turns lead nitrate paper black;

3 a gas which turns red litmus blue;

4 a gas which forms dense white fumes with ammonia;

5 two gases that relight a glowing splint.

Questions 6–10. Choose a substance from the following list to answer each question:
Calcium nitrate, sodium nitrate, ammonium nitrate, copper(II) nitrate, aluminium nitrate, zinc nitrate.

Which substance:

6 gives an orange flame test?

7 evolves ammonia when heated with sodium hydroxide solution?

8 forms a blue precipitate when ammonia solution is added to a solution of the compound?

9 forms a white precipitate when ammonia solution is added, and this precipitate does not redissolve?

10 forms a white precipitate when ammonia solution is added, and this precipitate redissolves in excess?

Questions 11–15. Choose a substance from the following list to answer each question:
sodium chloride, sodium nitrate, sodium sulphate, sodium carbonate, sodium sulphide, sodium sulphite.

Which substance:

11 evolves a greenish-yellow gas when heated with a mixture of manganese(IV) oxide and concentrated sulphuric acid?

12 forms a white precipitate when dilute hydrochloric acid and barium chloride solution are added to a solution?

13 gives a positive brown ring test?

14 evolves no gas in the cold with dilute hydrochloric acid but evolves a pungent gas on warming?

15 forms a white precipitate when dilute nitric acid and silver nitrate solution are added to a solution?

Test on Unit 42 The diaphragm cell

Questions 1–3. Which gas is produced in the diaphragm cell

1 at the cathode;

2 at the anode;

3 at the anode if the concentration of sodium chloride in the solution is low?

4 Give one advantage of the diaphragm cell compared to the Kellner-Solvay cell. (Unit 14.3)

5 Give one disadvantage of the diaphragm cell compared to the Kellner-Solvay cell.

Questions 6–8. Complete the following flow diagram by including the names and formulae of the products.

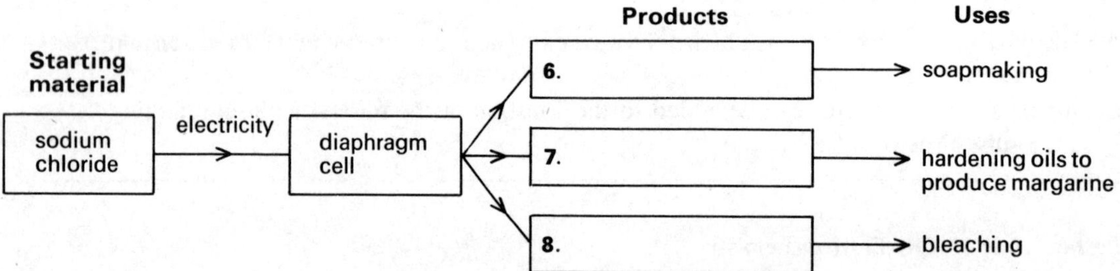

9 Why is the anode made of carbon rather than iron?

10 How could hydrochloric acid be produced easily from the products of the electrolysis?

Answers to self-test units

Unit 1

1 Element
2 Mercury
3 Helium or chlorine
4 Sodium, mercury
5 Synthesis
6 Sodium oxide
Sodium sulphide
7 Sodium oxide
Sodium sulphite
Sodium thiosulphate
Sodium sulphate
Sodium hydroxide
8 False
9 Potassium, carbon, nitrogen, oxygen
10 KCNS
11 Potassium nitrite
12 Magnesium nitride
13 Calcium, hydrogen, oxygen
14 Sodium, hydrogen, sulphur, oxygen
15 Sodium, hydrogen

Unit 2

1 Dissolves
2 Solution
3 Insoluble
4 Filtration
5 Evaporation (or boiling)
6 Distillation
7 Decantation (or filtration)
8 Filtration
9 Distillation
10 Fractional distillation
11 Sublimation
12 Fractional distillation
13 Chromatography
14 Using a tap funnel
15 Filtration

Unit 3

1 Sodium hydroxide
2 Nitric acid
3 Sulphuric acid
4 Calcium
Hydrogencarbonate
5 Magnesium nitrate
6 Zinc hydroxide
7 Ammonium carbonate
8 AgCl
9 PbS
10 K_2CO_3
11 $FeCl_2$
12 $FeCl_3$
13 $Fe_2(SO_4)_3$
14 Na_2SO_3
15 $Ba_3(PO_4)_2$
16 2HCl
17 2HCl
18 $2H_2O$
19 2Na, $2H_2O$, 2NaOH
20 2NaCl

Unit 4

1 Electron
2 Proton
3 Neutron
4 Electron
5 Electron
6 Isotopes
7 11 protons
12 neutrons
11 electrons
8 11 protons
12 neutrons
10 electrons
9 Li 2,1
Na 2,8,1
K 2,8,8,1
10 One electron in outer energy level
11 Protons and electrons
12 Atom
13 True
14 False
15 Only 1 isotope

Unit 5

1 Allotropy
2 Magnesium oxide (or calcium oxide)
3 Ionic
4 Yes
5 No
6 No
7 Yes
8 True
9 True
10 Sharing, transfer
11 $Mg \rightarrow Mg^{2} + 2e^-$
12 $F + e^- \rightarrow F^-$
13 Ionic
14 Solid (high melting point)
15 MgF_2

Unit 6

1 Liquid
2 Gas
3 Boiling point
4 Liquid
5 Solid
6 Freezing point (or melting point)
7 Gas
8 Solid
9 Gas
10 Diffusion

Unit 7

1 Decreases
2 Boyle's
3 Increases
4 Charles'
5 0°C (273K)
6 1 atmosphere (760 mm)
7 296K
8 54.6 cm^3
9 28.3 cm^3
10 False

Unit 8

1 Nitrogen
2 Nitrogen
3 Oxygen, water
4 Photosynthesis
5 Carbon dioxide, water vapour
6 Iron (III) oxide
7 Greasing, painting, galvanising
8 Oxygen
9 64 cm^3
10 Copper(II) oxide
11 To allow the gas to contract
12 True
13 Carbon monoxide is produced from incomplete combustion of petrol
14 Combustion (burning or oxidation)
15 False

Unit 9

1 Carbonic acid H_2CO_3
2 Nitrogen monoxide NO
3 Red lead Pb_3O_4
4 Sulphur dioxide SO_2
5 Sodium zincate Na_2ZnO_2
6 False (see Unit 9.2)
7 Neutral; Alkaline
8 Acidic
9 Amphoteric
10 Hydrogen peroxide

Unit 10

1 Zinc (or magnesium or iron)
2 Zinc or aluminium
3 Concentrated sulphuric acid
4 Water
5 Copper
6 Mixtures of hydrogen and air can explode
7 Copper
8 Water
9 To prevent air reacting with hot copper
10 Lead(II) oxide
Iron(III) oxide

Unit 11

1 Calcium stearate
2 Calcium sulphate
3 Calcium hydroxide
4 Calcium hydrogen carbonate
5 Anhydrous calcium chloride
6 True
7 Soapless detergent
8 *(i)* Sodium chloride
(ii) Potassium nitrate
9 False. It only proves that it *contains* water
10 False

Unit 12

1 False. It contains 2 moles of hydrogen atoms
2 0.8 g
3 8 g
4 Cu_2O
5 100 g
6 48%
7 1 g
8 2
9 0.25
10 SO_3
11 250 g
12 32 g
13 24,000 cm^3
14 1000 cm^3
15 83.3 cm^3

Unit 13

1 False – because carbon dioxide escapes

2 2 : 1
3 Law of multiple proportions
4 True
5 False – law of constant composition
6 80 cm^3
7 20 cm^3
8 60 cm^3
9 80 cm^3
10 20 cm^3 – water condenses and only nitrogen remains

Unit 14

1 Brine (sodium chloride solution)
2 Mercury
3 Amalgam
4 Hydrogen
5 Chlorine
6 Lead
7 Mercury
8 Lead
9 Aluminium
10 Lead
11 Negative
12 Positive
13 Negative
14 Increases
15 Alloy

Unit 15

1 Magnesium
Aluminium
Iron
Lead
Copper
Silver
2 Mercury
3 Potassium, calcium
4 Potassium
5 Copper, mercury
6 Zinc sulphate, copper
7 Solution turns colourless
Brown copper deposited
8 Molten
9 Lower
10 Negative
11 Positive
12 Blasts of hot air
13 and 14 Coke, limestone
15 Carbon monoxide

Unit 16

	Oxidised	*Reduced*
1	Aluminium	Iron(III) oxide
2	Iodine ion	Chlorine
3	Magnesium	Hydrogen ion
4	Magnesium	Copper(II) oxide
5	Zinc	Copper(II) ion
6	Carbon	Carbon dioxide
7	Aluminium	
8	Iodine ion	
9	False	
10	False	

Unit 17

1 Lattice
2 Free ions
3 Electrolyte
4 Electrolysis
5 Iodine
6 Anode
7 Potassium
8 Cathode
9 Iodine
10 Hydrogen
11 $K^+(aq)$, $H^+(aq)$
$I^-(aq)$, $OH^-(aq)$
12 965 coulombs
13 0.01 Faradays
14 0.2 g
15 0.8 g

Unit 18

1 Lithium
2 Oil
3 Lithium oxide
4 Lithium hydroxide and hydrogen
5 Iodine
6 Iron(III) bromide
7 *(i)*
8 D,E
9 A
10 C
11 B
12 E
13 C
14 A
15 Ionic

Unit 19

1 Sodium hydrogencarbonate
2 Calcium hydroxide
3 Silver chloride, zinc carbonate
4 Sulphuric acid; water
5 Calcium carbonate; water, carbon dioxide
6 Zinc, hydrochloric acid
7 Barium sulphate, sodium nitrate
8 7
9 True
10 Lead nitrate
11 To speed up the reaction or to prevent copper(II) crystals forming
12 To ensure all acid is used
13 To remove excess copper(II) oxide
14 Crystals may decompose
15 $CuO(s) + H_2SO_4(aq) \rightarrow CuSO_4(aq) + H_2O(l)$

Unit 20

1 Hydrogen
$Mg(s) + 2HCl(aq) \rightarrow MgCl_2(aq) + H_2(g)$
2 *(i)* 90 cm^3
(ii) 50 cm^3
(iii) 32 cm^3
3 As reaction proceeds acid becomes less concentrated
4 Reaction has stopped
5 *(i)* Increased
(ii) Decreased
(iii) Increased
6 *(i)* Increased
(ii) Decreased
(iii) Increased
(iv) Unchanged

Unit 21

1 Precipitate
2 To the right
3 Add concentrated hydrochloric acid
4 True
5 To the right
6 To the right
7 Unchanged
8 To the right
9 Moves to the right
10 Moves to the left

Unit 22

1 Plastic sulphur
2 α-sulphur
3 α-sulphur
4 Plastic sulphur
5 Iron(II) sulphide
6 Frasch
7 Water
8 Outer
9 Melt
10 Air

Unit 23

1 Zinc sulphide
2 $ZnSO_4(aq) + H_2S(g) \rightarrow ZnS(s) + H_2SO_4(aq)$
3 Hydrogen sulphide
4 Sulphides
5 True
6 Sulphur dioxide
7 Iron(II) chloride, sulphur, hydrochloric acid
8 Sulphur, hydrogen chloride
9 Lead(II) sulphide, nitric acid
10 Sulphur and water (see Unit 16.6)

Unit 24

1 A: Hydrogen sulphide
2 B: Sulphur trioxide (sulphur(VI) oxide)
3 C: Sulphuric acid
4 D: Copper(II) sulphate
5 E: Sulphur
6 F: Calcium sulphite
7 G: Sulphur dioxide
8 False
9 True
10 False

Unit 25

1 A: Carbon
2 B: Hydrogen chloride
3 C: Sulphur dioxide
4 D: Iron(III) oxide
5 E: Sulphuric acid
6 Sulphur, zinc sulphide, iron pyrites anhydrite or hydrogen sulphide
7 Impurities poison catalyst
8 $2SO_2(g) + O_2(g) \rightleftharpoons 2SO_3(g)$
9 Vanadium(V) oxide
10 Oleum, Water

Unit 26

1 Potassium hydroxide
2 Copper
3 Noble
4 and 5 Sodium nitrite and ammonium chloride
6 Ammonium nitrate
7 Water
8 Lightning
9 Bacteria
10 $3Mg(s) + N_2(g) \rightarrow Mg_3N_2(s)$

Unit 27

1 A: Ammonia solution or ammonium hydroxide
2 B: Water
3 C: Ammonium chloride
4 D: Ammonia
5 E: Iron(III) hydroxide
6 False
7 True
8 Haber
9 Hydrogen
10 Iron

Unit 28

1 Concentrated sulphuric acid
2 Yellow
3 Air
4 A: Magnesium
5 B: Zinc nitrate
6 C: Copper(II) oxide
7 D: Concentrated nitric acid
8 E: Potassium or sodium nitrite
9 True
10 True

Unit 29

1 Dinitrogen monoxide
2 Dinitrogen monoxide
3 Nitrogen monoxide
4 Nitrogen dioxide
5 Dinitrogen monoxide
6 Nitrogen dioxide
7 Nitrogen dioxide
8 Nitrogen dioxide
9 Nitrogen monoxide
10 Nitrogen monoxide

Unit 30

1 Nitrogen, phosphorus, potassium, magnesium, calcium, sulphur
2 Trace
3 Manure, dried blood, sodium nitrate
4 Soluble
5 It is insoluble
6 Ammonia
7 Nitrates
8 Nitrogen
9 Potassium
10 Phosphorus

Unit 31

1 Calcium phosphate, sand, coke
2 False
3 Phosphorus(V) oxide
4 Phosphoric acid
5 Acidic
6 False
7 Under water
8 White phosphorus
9 White
10 Phosphates

Unit 32

1 Concentrated
2 Sodium chloride
3 Chlorine
4 Oxidises
5 Sodium chloride, water, sodium chlorate(I)
6 Carbon, hydrogen chloride
7 Iron(II) chloride, hydrogen
8 Manganese(II) chloride, chlorine, water
9 Sulphur, hydrogen chloride
10 $2HOCl(aq) \rightarrow 2HCl(aq) + O_2(g)$

Unit 33

Part A Carbon

1 True
2 $CO_2(g) + C(s) \rightarrow 2CO(g)$
3 Passing through limewater or potassium hydroxide solution
4 $Fe_2O_3(s) + 3CO(g) \rightarrow 2Fe(s) + 3CO_2(g)$
5 Carbon dioxide

Part B Carbon monoxide

1 Figs. 33.1 and 2
2 True
3 $CuO(s) + C(s) \rightarrow Cu(s) + CO(g)$
4 Reducing agent
5 True

Part C Organic chemistry

1 Fractional distillation
2 Cracking
3 Ethene
4 Iodine solution
5 Fehling's (or Benedict's) solution
6 Bromine solution
7 E
8 B,C
9 A,D
10 B

Unit 34

1 Marble, limestone, chalk, shells
2 Calcium oxide
3 Calcium hydroxide (limewater)
4 A: Sodium hydrogen carbonate
5 B: Copper carbonate
6 C: Sodium hydrogen carbonate
7 D: Calcium carbonate
8 E: Carbon
9 Sodium Hydrogencarbonate, $NaHCO_3$
10 8.0

Unit 35

1 Condensation
2 Addition
3 Bakelite
4 Polythene
5 Double
6 True
7 True
8 Ethene
9 False
10 True

Unit 36

1 False
2 Carbon dioxide
3 Coke, coal gas, coal tar, ammonia solution
4 Destructive distillation
5 By the decay of plant and vegetable material
6 Methane CH_4
7 Producer gas; carbon monoxide, nitrogen
8 Water gas; carbon monoxide, hydrogen
9 In 7 only one gas can burn
10 Hydrogen. Greater heat of combustion per gram

Unit 37

1 4.2 kJ
2 16.8 kJ
3 Heat losses
4 False
5 Chemical; electrical
6 1967.5kJ
7 Formation
8 – 34 kJ/mole
9 Exothermic
10 –86 kJ

Unit 38

1 North Sea supply of natural gas
2 Sulphur dioxide
3 Short-chain hydrocarbons are easier to sell
4 Ammonia (from ammonium chloride), carbon dioxide
5 In the extraction of zinc (see unit 15.8)

Unit 39

1 γ-rays
2 β-particles
3 α-particles
4 β-particles
5 α-particles
6 128 cpm
7 32 cpm
8 44 cpm
9 Atomic number 90; Mass number 227
10 Atomic number 82; Mass number 210

Unit 40

1 4 g
2 0.0025 moles
3 0.0025 moles
4 31.25 cm^3
5 False

Unit 41

1 Nitrogen dioxide
2 Hydrogen sulphide
3 Ammonia
4 Hydrogen chloride
5 Oxygen, dinitrogen monoxide
6 Sodium nitrate
7 Ammonium nitrate
8 Copper(II) nitrate
9 Aluminium nitrate
10 Zinc nitrate
11 Sodium chloride
12 Sodium sulphate
13 Sodium nitrate
14 Sodium sulphite
15 Sodium chloride

Unit 42

1 Hydrogen
2 Chlorine
3 Oxygen (from the discharge of OH^- ions)
4 Avoids the use of mercury which is expensive.
5 Sodium hydroxide produced is not pure.
6 Sodium hydroxide NaOH
7 Hydrogen H_2
8 Chlorine Cl_2
9 Chlorine produced might react with iron $2Fe + 3Cl_2 \rightarrow 2FeCl_3$.
10 Burning hydrogen in chlorine to produce hydrogen chloride. The hydrogen chloride is dissolved in water to produce hydrochloric acid.

Section IV Hints for candidates taking chemistry examinations

Having prepared yourself thoroughly, only poor examination technique can prevent you doing justice to yourself. The following points are worth remembering.

1 The importance of reading the questions thoroughly and carefully

It is frequently stated in reports by examiners that candidates misread questions or fail to use the information given in the question. If the question concerns the industrial preparation of nitric acid, there could be no credit for a candidate whose answer refers to the laboratory preparation of nitric acid. If information (*e.g.* a table comparing two allotropes, an equation, relative atomic masses, a graph *etc.*) is given in a question, it must be required to answer the question fully.

The question 'Explain what you would SEE if excess iron filings are added to copper(II) sulphate solution' requires more than a correct equation. You would be expected to mention that copper(II) sulphate solution is blue and when excess iron filings are added, the blue solution goes colourless and a brown solid (copper) is deposited. It is obvious that most of the marks are awarded for these observations.

Where there is a choice of questions to be made, read all the questions through carefully before making any choice.

2 Spending too much time on one question or on part of the examination

A candidate who completes only half of the questions required can only achieve a maximum of 50% and can have little chance of being successful. It is important to divide your time equally between the questions. It is a good idea to prepare a timetable for the examination before entering the examination room and then to stick to it.

For example, if you are preparing to take a 2½ hour examination paper consisting of 40 fixed response questions and four longer questions to be chosen from eight, your plan might be:

1.30 pm. Start the fixed response questions. There is no point in reading all the fixed response questions before you start as they are compulsory.
1.55 pm. You should have reached item 20.
2.30 pm. Complete fixed response questions. Note that more time is allowed for later items in the fixed response section. Read the eight longer questions and select the four to be attempted.
2.40 pm. Attempt the question you feel you can do best. This is important as it will increase your confidence and will benefit you if, for any reason, you cannot complete the paper. If you have not finished at 3.00 pm, stop writing and move onto the next question.
3.00 pm. Start the second long question of your choice.
3.20 pm. Start the third long question of your choice.
3.40 pm. Start the fourth long question of your choice.
4.00 pm. Examination finishes.

If you have any time left, go back and complete any unfinished question.

It is unwise to abandon a question, when you have spent some time on it, in order to start another question. You will be wasting time with no certainty that you can do better with the other question.

3 Doing the wrong number of questions

Despite clearly stating the number of questions to be attempted, an examiner sees many papers where candidates have attempted the wrong number of questions.

It is believed by some candidates that if too many questions are attempted, the examiner will mark all the questions and credit the candidate with marks from the best answered questions. This is not so and the examiner will award marks for the first questions attempted. It can never benefit the candidate to do more questions than required. If you find you have time to spare that enables you to attempt other questions, your answers may not be sufficiently detailed.

If you have done too many questions, make sure you have crossed out thoroughly any question that you do not want to be marked.

4 Lack of planning and poor presentation

It is important to plan your answers carefully. This becomes more important as the questions get longer and more involved. A plan takes only a few minutes and it will help you to ensure that you have covered the full extent of the answer required.

Some candidates believe that there is some credit given for long answers. When marking a longer question an examiner is looking for the inclusion of certain facts or statements. Marks are awarded when these are included. Unless these long answers contain the required facts or statements, no marks can be awarded.

You can lose marks for bad presentation and untidy work. Examiners cannot award marks for answers they cannot read.

5 Chemical equations

Your answers to questions in Chemistry should include equations whenever relevant. If you are in doubt about the relevance of an equation, include it in your answer.

An equation is a useful summary of a chemical reaction. If three marks are awarded for an equation, one mark will be awarded for the correct word equation, one mark for the correct formulae throughout the equation and one mark for correctly balancing the equation. You are advised to include, therefore, both word and symbol equations.

For CSE, I believe candidates should concentrate on writing correct word equations.

If the question asks for a test for carbon dioxide, an equation (see unit 34.3) should be included for the reaction between limewater and carbon dioxide.

Check that the symbol equation is balanced before you move on. If it is impossible to balance the equation completely, it suggests that you may have missed one or more of the reactants or products or that you may have written one or more of the formulae incorrectly.

State symbols, *e.g.* (s), (l), (g) or (aq), are used in this book. They are a useful addition to your equations but they are not expected by all Examination Boards.

6 Chemical calculations

All Chemistry papers at 'O' level, or SCE or CSE contain chemical calculations. They are usually not very well attempted by candidates. The figures are chosen to minimise arithmetic and it should not be necessary to resort to mathematical tables, slide rules or calculators.

On many candidates' papers only answers are given and the working is not shown. If the answer is incorrect the examiner cannot award any marks if no working is shown. If the working is given, despite the wrong answer, it may still be possible to award a good mark. Remember most of the marks will be awarded for the essential *Chemistry* rather than incidental arithmetic.

Remember to give units, where appropriate, to your answer. Before moving on to the next question, check that your answer is reasonable. I have seen a candidate, after making an arithmetical mistake, write on an 'O' level paper 'because the ratio 41977:28493 is a simple ratio, the law of multiple proportions is verified'. If he had realised that this ratio is not simple, he might have looked back and found the mistake.

7 Attention to detail

One of the distinguishing features between a good candidate and an average candidate is the ability to incorporate details into the answer. As a guide, in any answer, you should include:

(*i*) Names of chemicals used and produced. Include states, colours and concentrations of the chemicals.
(*ii*) Give the conditions of any reaction, *i.e.* temperature, catalysts *etc.*
(*iii*) Explain why the reaction takes place. Relate your answer, if possible, to the Periodic Table or the reactivity series.
(*iv*) Write the equations in words and symbols.

8 Drawing diagrams

Diagrams should be included whenever they improve your answer. They should help you avoid having to write a long descriptive account of your experiment.

Do not spend a long time doing an artistic diagram. The important feature of your diagrams must be clarity. Draw the diagrams about half as big again as the diagrams in this book. Draw your diagrams freehand with a pencil and have an eraser available in case you make any mistakes. You may use stencils but they restrict your diagrams in size. When the diagram is finished, look carefully to make sure you have not made any obvious mistakes, *e.g.* in a gas preparation, a thistle funnel must enter the solution in the flask or the gas will escape. Finally, label every piece of apparatus and all chemicals in ink.

The following points are worth remembering:

(*i*) Do not waste time drawing stands and clamps. The examiner assumes you will support the apparatus correctly. They also detract from the important features of the diagram.

(*ii*) Do not waste time drawing Bunsen burners for heating part of your apparatus. Just draw an arrow and label it HEAT.

(*iii*) Try to draw each piece of apparatus to the correct size in relation to the other pieces of apparatus.

(*iv*) Draw a round-bottomed flask if heating is required and a flat-bottomed or conical flask if it is not. If a round-bottomed flask is used, it must be drawn above the level of other apparatus to enable heat to be applied (see Fig. 24.1).

(*v*) A common mistake in diagrams is to omit corks and bungs.

(*vi*) If a gas is to be bubbled through a liquid, ensure that the tube goes below the level of the liquid in the wash bottle, and the outlet for the gas is above the level of the liquid (see Fig. 24.1).

(*vii*) When collecting a gas in a gas jar using a beehive shelf (*e.g.* Fig. 9.1), ensure that the level of the water in the trough is above the top of the beehive shelf, that there is water inside the beehive shelf and that there is a hole in the top of the beehive shelf through which the gas can pass.

(*viii*) When collecting a gas by upward or downward delivery (*e.g.* Fig. 27.1) or Fig. 24.1), ensure that the delivery tube reaches to the end of the gas jar.

(*ix*) Having drawn an arrangement to dry a gas, do not collect it over water.

TYPES OF EXAMINATION QUESTION

Fixed response questions

Most syllabuses contain some fixed response questions. These are sometimes called multiple choice or objective questions. At first sight, these questions seem to be comparatively easy because it is simply a matter of choosing the correct answer from the possible answers given. In practice candidates do not always do as well as they expect, because the questions are specially designed, written and tested before use and because the candidates are limited in the time available.

Because these questions can be easily and accurately marked by hand or machine and because the standard of the examination can be judged before the examination by pre-testing, fixed response questions are popular with Examination Boards.

There are four types of fixed response question frequently used in 'O' level and CSE examinations.

Type 1 Simple multiple choice

In this type of item you are given an incomplete statement or question (called the stem) together with four or five possible answers (or responses) and you have to choose the only one that correctly fits the stem. The wrong answers are called distractors.

E.g. What colour is hydrated copper(II) sulphate?

A. blue
B. green
C. white
D. yellow
E. black

A is the correct answer (or key). B, C, D and E are distractors. This question is entirely recall of information that the candidate should have learnt. If he confuses hydrated copper(II) sulphate with anhydrous copper(II) sulphate, he would answer C.

Many questions of this type, particularly on 'traditional' syllabuses, test factual learning in this way. If you have fully understood the units in Section II in this book, these questions should not be too difficult.

Other questions of this type require, in addition to factual learning, understanding and/or the application of principles.

E.g. Which one of the italicised substances is an element?

A. *A green substance* which is separated into two substances by chromatography.
B. *A black liquid* which boils over a wide range of temperature.
C. *A black solid* which burns in oxygen completely to form a single colourless gas.
D. *A white substance* which turns yellow on heating but turns white again on cooling.
E. *A colourless liquid* which turns to a colourless solid on cooling.

Any candidate who has just learnt the definition of the term 'element' but does not understand it, is unable to answer this question. The correct response is C because the black solid burns completely to form only one product. There is no need to try to identify these substances.

Sometimes a number of these items may be linked to the same experimental situation.

Type 2 Classification

This type of question is used when a number of similar Type 1 fixed response questions are being set with identical responses. The five lettered responses are given first followed by the series of questions.

E.g. For each of the questions 1–4 choose the one process labelled A, B, C, D or E with which it is chiefly associated.

A. cracking
B. polymerisation
C. oxidation
D. hydrogenation
E. neutralisation

1 Decane vapour is changed to ethene when passed over heated china.

2 Margarine is produced by passing hydrogen through a heated oil in the presence of a catalyst.

3 Sodium carbonate is added to ethanoic acid (acetic acid) until no further carbon dioxide is evolved.

4 Ethene is completely burnt in excess oxygen.

The correct responses to these questions are:

1 A, **2** D, **3** E, **4** C

In this type of question each response may be used once, more than once or not at all.

Type 3 Multiple completion questions

In this type of question there is more variation between Examination Boards. Basically, a stem is given together with four responses (numbered 1, 2, 3 and 4). However, one or more than one of these responses is (are) correct. Different Examination Boards use different combinations of responses. A typical combination for 'O' level is:

choose A. if only 1, 2 and 3 are correct
B. if only 1 and 3 are correct
C. if only 2 and 4 are correct
D. if only 4 is correct
E. if some other response, or combination of responses is (are) correct

These directions are summarised in Table 1.

Table 1.

A	B	C	D	E
1, 2, 3 only	1, 3 only	2, 4 only	4 only	Some other response or combination of responses

Example:
A heavy (or transition) metal such as manganese

1. conducts electricity
2. burns to form an acidic oxide
3. forms a wide range of coloured compounds
4. is placed in Group 1 of the Periodic Table

When attempting this type of question, it is worth remembering that it is possible to arrive at answer E by completely incorrect reasoning. If the correct combination should be 2 and 3, your correct response should be E. If, by incorrect reasoning, you believed the correct answer was 1 and 4, your response again would be E. In other words, you would be getting a mark for totally incorrect reasoning. For this reason, the examiner does not use the response E very often in this type of question.

In the example above, manganese is a metal and therefore conducts electricity—response 1 is correct. Response 2 is incorrect because metals form neutral or alkaline oxides. Response 3 is correct and response 4 is incorrect. The correct answer (by reference to the table) is B (1 and 3 correct only). The fact that the answer fits one of the four combinations (A, B, C or D) should give the candidate some satisfaction.

This type of question is regarded as too difficult by some CSE Examination Boards. It is either not included or included in a simpler form. For example only three responses (1, 2 and 3) may be given and the combinations simplified as follows:

choose A. if 1, 2 and 3 are correct
B. if only 1 and 2 are correct
C. if only 2 and 3 are correct
D. if only 1 is correct
E. if only 3 is correct

Type 4 Assertion-reason questions

True or false questions appear on some CSE papers but as there are only two possible responses (true or false), the marking has to allow for guessing. Assertion-reason questions are really a development of true or false questions. First you have to decide whether the first statement is true or false and then decide if the second statement is true or false.

If both statements are true and the second statement is a correct explanation of the first statement, you should answer A. If both statements are true but the second statement is *not* a correct explanation of the first, the answer is B. If the first statement is true but the second statement is false the answer is C. If the first statement is false and the second statement is true, the answer is D. If both statements are false, the answer is E. These instructions are summarised in Table 2.

Table 2.

	First statement	*Second statement*	
A	True	True	Second statement is a correct explanation of the first.
B	True	True	Second statement is NOT a correct explanation of the first.
C	True	False	
D	False	True	
E	False	False	

E.g. First statement
A luminous (yellow) Bunsen burner flame is cooler than a non-luminous (blue) flame.

Second statement
The gas is not completely burnt in the luminous flame.

Both statements are correct and so response A or B must be correct. The correct answer is A because the second statement is a correct explanation of the first. If the correct answer is A or B you have to do one extra step and so answers A and B are used more frequently in this type of question.

This type of question is not widely used on CSE papers.

Points to be remembered when attempting a fixed response test:

1. Read through each question correctly. Often a candidate gives a wrong answer because the question has not been read and understood.
2. Do not spend too much time on the early items or on any single item. Invariably the questions get longer and more involved as the test progresses, and you will need more time to tackle items near the end of the test.
3. Because there are a large number of questions in each test, it is unwise to study only parts of the syllabus. You should attempt to understand as much of the syllabus as possible.
4. Do not be afraid to guess. Guessing sensibly can help you and you are expected to do it. If you cannot answer a question at all, read all the responses and then guess. If there are five possible responses and you know that three are wrong but you cannot decide between the other two, then guess. You have increased your chances of success by ruling out incorrect responses.
5. Make sure you use all the information given.
6. Research has shown that in fixed response tests repeated checking of your answers does not improve the final mark obtained. The allocation of time does not allow for time to check answers at the end of the examination.

See page 178 for examples of fixed response questions.

Free response questions

Free response questions can be defined as any question where the candidate has to make a written answer rather than just choose from the answers given. The written answer may be a **word**, a **phrase**, a **sentence**, or an **essay**.

There is a wide range of free response questions ranging from short answer questions (where the answer may be just one word), through structured questions (where the question is divided into a number of parts all relating to the same situation) to essay-type questions. The distinction between essay-type questions and structured questions is not very clear. Many examinations include essay questions that are structured to assist the candidates.

It is important to read the question through carefully before choosing to do a question. If information is given in the question it should be used in answering the question. Have regard for the amount of space given on the paper for your answer and for the number of marks awarded for each part. Remember that the examiner is only human and answer the question clearly.

See page 181 for examples of free response questions.

REASSURANCE

If you have worked conscientiously through the relevant sections of this book you will be well prepared for the examination. In addition to having the knowledge required, you will be aware of common mistakes that candidates make and how to ensure that you produce your best answers.

There are two things you should remember as you prepare to take this examination. For 'O' level, you have to obtain only about 45% to achieve a grade C. In other words, you have to get less than half of what you should attempt right. Also, of the candidates who take 'O' level Chemistry, over 70% of them achieve a grade C or better. Having prepared yourself thoroughly, you should be among the successful.

Section V Practice in answering examination questions

The following multiple choice questions have been compiled by the author and are based on similar questions set by GCE and CSE Boards. The answers may be found on page 181.

MULTIPLE CHOICE QUESTIONS

1 Magnesium (atomic number 12) forms Mg^{2+} ions. The number of electrons in a magnesium ion is

A. 4
B. 6
C. 10
D. 12
E. 14

2 Which one of the following substances dissolves in water to form a solution with a pH greater than 7?

A. Sulphur dioxide
B. Ammonia
C. Chlorine
D. Copper(II) oxide
E. Hydrogen chloride

3 Which one of the following reactions would result in the formation of an element?

A. Burning carbon in excess air.
B. Heating copper(II) sulphate crystals.
C. Neutralising an acid with a base.
D. Reducing lead(II) oxide with hydrogen.
E. Heating lead(II) carbonate.

4 Dehydration of methanoic acid (formic acid) HCOOH produces a gas. This gas is

A. carbon monoxide
B. carbon dioxide
C. hydrogen
D. carbon
E. water.

5 An indicator is used during a neutralisation reaction in order to

A. detect the acid and the alkali.
B. show when exactly reacting quantities of acid and alkali are present.
C. speed up the rate of reaction between the acid and the alkali.
D. measure the amount of heat liberated.
E. show whether the reaction is reversible.

6 Crystals of sodium carbonate decahydrate (washing soda) are efflorescent. When these crystals are exposed to air the crystals

A. lose mass and remain solid.
B. gain mass and remain solid.
C. gain mass and become liquid.
D. gain mass, change to a liquid and evolve bubbles of gas.
E. remain unchanged.

7 When a piece of copper is added to silver nitrate solution, silver is displaced. Iron reacts slowly with warm, dilute hydrochloric acid to produce hydrogen but silver and copper do not react. The metals in order of reactivity, with the most reactive first, are

A. copper, silver, iron
B. iron, silver, copper
C. iron, copper, silver
D. copper, iron, silver
E. silver, iron, copper. ☐

8 Crude oil is separated into fractions with different boiling points by fractional distillation. In which one of the following are the fractions arranged in the correct order of increasing boiling point?

A. Petrol, bitumen, diesel oil, paraffin
B. Bitumen, diesel oil, paraffin, petrol
C. Diesel oil, petrol, bitumen, paraffin
D. Paraffin, diesel oil, bitumen, petrol
E. Petrol, paraffin, diesel oil, bitumen ☐

9 2.38 g of tin, when treated with concentrated nitric acid and heated, produced 3.02 g of anhydrous tin oxide. What is the formula of this oxide? (Sn = 119, O = 16)

A. SnO
B. SnO_2
C. Sn_2O
D. Sn_2O_3
E. SnO_3 ☐

10 $2NH_3(g) + 3CuO(s) \rightarrow N_2(g) + 3H_2O(g) + 3Cu(s)$

What is the volume of the gaseous product formed when 80 cm^3 of ammonia is passed over heated copper(II) oxide?
(All volumes measured at room temperature and pressure.)

A. 20 cm^3
B. 40 cm^3
C. 80 cm^3
D. 120 cm^3
E. 160 cm^3 ☐

Questions 11–15 refer to the following types of chemical reaction:

A. Hydrolysis
B. Polymerisation
C. Dehydration
D. Neutralisation
E. Precipitation

Select the term from the list above which best describes the reaction represented by each of the following equations:

11 $(C_6H_{10}O_5)_n(aq) + n\ H_2O(l) \rightarrow n\ C_6H_{12}O_6(aq)$ ☐

12 $OH^-(aq) + H^+(aq) \rightarrow H_2O(l)$ ☐

13 $C_2H_5OH(g) \rightarrow C_2H_4(g) + H_2O(g)$ ☐

14 $MgO(s) + 2HNO_3(aq) \rightarrow Mg(NO_3)_2(aq) + H_2O(l)$ ☐

15 $MgSO_4(aq) + Na_2CO_3(aq) \rightarrow MgCO_3(s) + Na_2SO_4(aq)$ ☐

Questions 16–20 refer to the curves labelled A–E. Each curve shows the change of property with time.

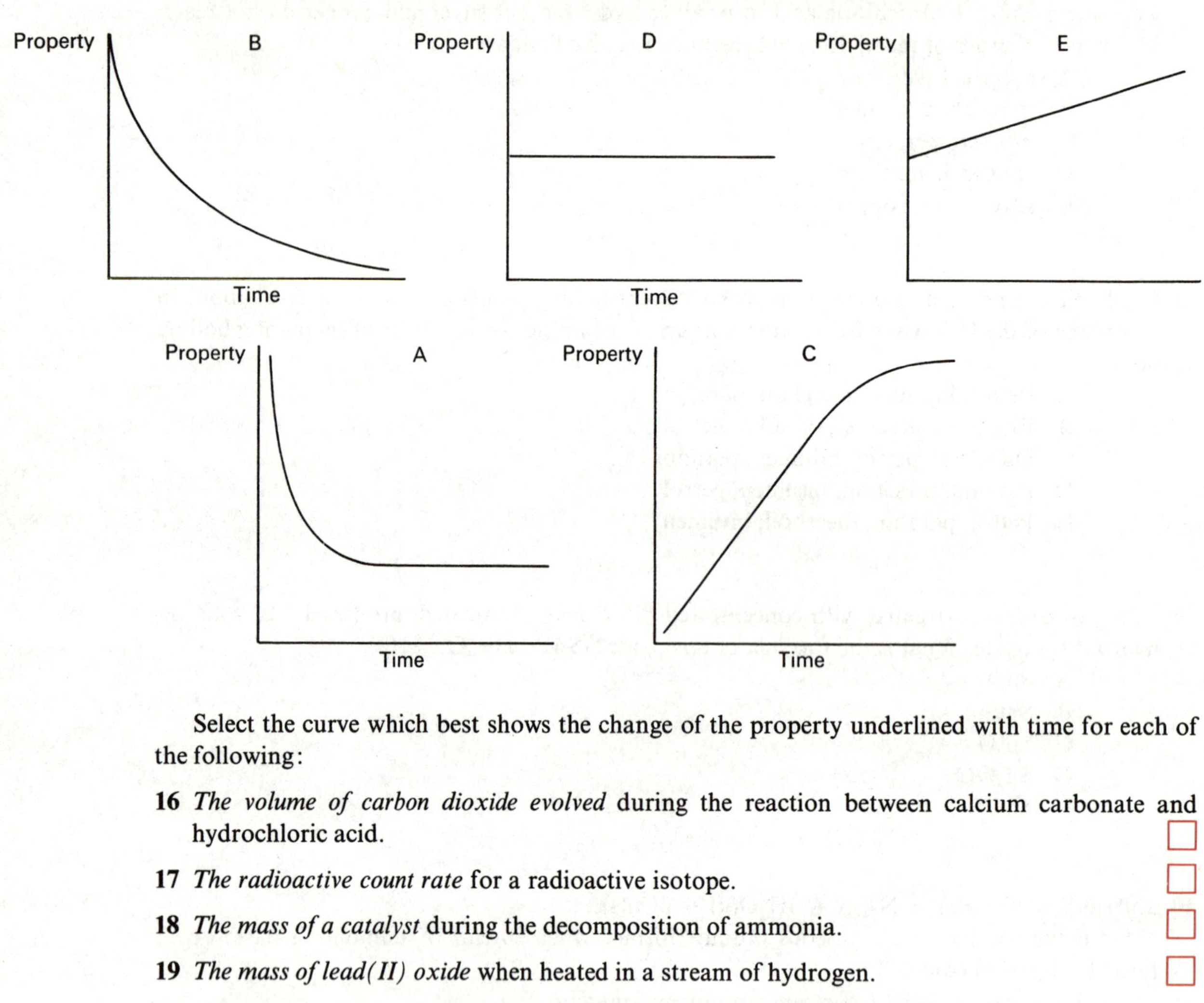

Select the curve which best shows the change of the property underlined with time for each of the following:

16 *The volume of carbon dioxide evolved* during the reaction between calcium carbonate and hydrochloric acid. ☐

17 *The radioactive count rate* for a radioactive isotope. ☐

18 *The mass of a catalyst* during the decomposition of ammonia. ☐

19 *The mass of lead(II) oxide* when heated in a stream of hydrogen. ☐

20 *The mass of a copper cathode* during the electrolysis of copper(II) sulphate solution with copper electrodes and a constant current. ☐

Questions 21–25.

For each of these questions one or more than one of the responses given is (are) correct. Then choose

A. if only 1, 2 and 3 are correct;
B. if only 1 and 3 are correct;
C. if only 2 and 4 are correct;
D. if only 4 is correct;
E. if some other response, or combination of responses is (are) correct.

21 Limestone is added to a blast furnace for the extraction of iron from iron ore in order to
1 lower the melting point of the iron.
2 produce additional carbon dioxide.
3 ensure that all the iron ore is used up.
4 remove silicon dioxide impurities from the furnace. ☐

22 Nitrogen dioxide (NO_2) gas can be produced by
1 heating copper(II) nitrate crystals.
2 heating sodium nitrate crystals.
3 the action of concentrated nitric acid on copper.
4 mixing two volumes of oxygen with one volume of nitrogen. ☐

23 Methods available for extracting metals from their compounds include
1 using a magnet.
2 electrolysis of the molten chloride.

3 heating the ore in a stream of chlorine.
4 reduction of ores with carbon.

24 Hard water can be softened by
1 passing through an ion-exchange column.
2 distillation.
3 adding sodium carbonate.
4 filtering.

25 Crystals of lead can be produced by
1 splitting open crystals of lead(II) nitrate.
2 adding a piece of zinc to a solution of lead(II) nitrate.
3 adding a piece of silver to a solution of lead(II) nitrate.
4 electrolysis of lead(II) nitrate solution.

Questions 26–30.

Each of the following questions consists of a statement in the left-hand column followed by a second statement in the right-hand column.

Decide whether the first statement is true or false.

Decide whether the second statement is true or false.

Then choose:

A. if both statements are true and the second statement is a *correct explanation* of the first statement.
B. if both statements are true but the second statement is NOT a *correct explanation* of the first statement.
C. if the first statement is true, but the second statement is false.
D. if the first statement is false, but the second statement is true.
E. if both statements are false.

	First statement	*Second statement*
26	Lead sulphate can be prepared by precipitation.	Lead sulphate is insoluble in water.
27	Molten lead bromide conducts electricity.	All substances that conduct electricity contain a metal.
28	Nitric acid is prepared in the laboratory in an all-glass apparatus.	Concentrated nitric acid is decomposed when strongly heated.
29	The initial rate of reaction between lumps of calcium carbonate and dilute hydrochloric acid can be increased by crushing the lumps into a powder.	Crushed calcium carbonate has a larger surface area than an equal mass of lumps of calcium carbonate.
30	Ethane C_2H_6 decolourises a solution of bromine.	Ethane is unsaturated.

Answers to multiple choice questions

1. C 2. B 3. D 4. A 5. B 6. A 7. C 8. E 9. B 10, B 11. A 12. D 13. C 14. D 15. E 16. C 17. B 18. D 19. A 20. E 21. C 22. B 23. C 24. A 25. C 26. A 27. C 28. B 29. A 30. E.

FREE RESPONSE QUESTIONS

The majority of the questions that follow have been selected from GCE, SCE and CSE Chemistry papers of a range of Examination Boards. The answers which are given to these questions are not the official answers from Examination Board marking schemes, but have been written by the author of this book. They are skeleton answers only.

GCE questions (answers on page 186)

The following questions are factual and the answers should consist of a single word or a phrase.

1 Give the name of

(*a*) a white refractory oxide;
(*b*) the colourless gas evolved when chlorine water is exposed to bright sunlight;
(*c*) a colourless gaseous compound which will rekindle a glowing spill;
(*d*) a solid acid salt which turns red litmus paper blue.

(Southern Universities Joint Board)

Questions 2–7 are structured questions and these are very common on all examination papers. They consist of a certain amount of information (possibly including tables or diagrams) at the beginning of the question followed by a series of short questions (or subquestions) related to the information given. These subquestions often get progressively more difficult. The information given clearly defines the limits of the question to the candidate. It is essential to read thoroughly and understand the information given. The candidate is also guided in the length of answer required by the amount of space provided (if the answer is to be fitted into spaces on the question paper) or by the maximum mark to be awarded (often shown in brackets after the question).

2 From the information given below, suggest the identities of **A**, **B**, **C** and **D**, and where possible explain the reactions described.

(*a*) The black powder, **A**, dissolves in dilute sulphuric acid without effervescence giving a blue solution. Addition of dilute ammonia solution causes the formation of a pale blue precipitate which dissolves to give a dark blue solution when excess ammonia solution is added. (5 lines, 6 marks)

(*b*) **B** is a solid element which ignites when touched with a warm glass rod and then burns vigorously. The dense white smoke given off settles as a powder which is soluble in water to give an acidic solution. (5 lines, 6 marks)

(*c*) The white solid, **C**, gives a brick red colour when a flame test is carried out. After exposure to the atmosphere for several days the solid appears to liquefy and this liquid gives a white precipitate with silver nitrate solution. (5 lines, 6 marks)

(*d*) **D** is a sodium salt. On heating in a test tube it melts, then liberates a gas which re-ignites a glowing splint and, on cooling, a white residue remains. When **D** is warmed with concentrated sulphuric acid, oily drops condense on the cooler parts of the tube. (5 lines, 6 marks)

(Associated Examining Board Syllabus I)

3 The following symbols refer to atoms of sodium, fluorine and neon.

$^{23}_{11}Na$ $\quad$ $^{19}_{9}F$ $\quad$ $^{20}_{10}Ne$ $\quad$ $^{22}_{10}Ne$

Using the above information, answer the following questions:

(*a*) What are the electronic structures of the sodium and fluorine atoms? (2 lines, 2 marks)

(*b*) Sodium and fluorine combine to give the ionic compound, sodium fluoride. Explain, with the aid of a diagram, the changes in electronic structures that take place in this reaction. (2 lines and diagram, 2 marks)

(*c*) State **two** characteristic properties that would be expected of the ionic compound, sodium fluoride. (2 lines, 2 marks)

(*d*) In what respects do the neon atom, the sodium ion and the fluoride ion
(*i*) resemble each other, (1 line, 1 mark)
(*ii*) differ from each other? (1 line, 1 mark)

(*e*) By means of a diagram, give the electronic structure of the covalent fluorine molecule, (F_2). (diagram, 2 marks)

(*f*) (*i*) What can you deduce about the composition of the neon atoms shown above? (2 lines, 2 marks)
(*ii*) How do you account for the fact that the relative atomic mass of naturally-occurring neon is 20.2? (2 lines, 3 marks)

(Welsh Joint Education Committee—GCE)

4 Describe and explain, giving **brief** practical details, the reactions by means of which **each** of the following conversions could be carried out in the laboratory:

(Each conversion requires more than one step.)

(*a*) lead(II) oxide → lead(II) sulphate, (6 lines, 7 marks)
(*b*) copper(II) carbonate → copper, (6 lines, 6 marks)
(*c*) calcium carbonate → carbon monoxide. (6 lines, 7 marks)

(Welsh Joint Education Committee—GCE)

5 (a) Two oxides of the element molybdenum (symbol Mo; relative atomic mass 96) gave the following results on analysis:
oxide **A**: 1.92 g of molybdenum were combined with 0.64 g of oxygen;
oxide **B**: 2.00 g of molybdenum were combined with 1.00 g of oxygen.
(*i*) Calculate the mass of oxygen combined with one mole of molybdenum in each oxide. (2 lines, 2 marks)
(*ii*) Show how these results illustrate an important law of Chemistry (3 lines, 2 marks)
(*iii*) Deduce the formulae of the oxides **A** and **B**. (5 lines, 4 marks)

(*b*) Describe how the following conversions may be carried out in the laboratory:
(*i*) sulphur dioxide → sulphur(VI) oxide, SO_3 (5 lines, 6 marks)
(*ii*) carbon dioxide → carbon monoxide (5 lines, 6 marks)

(Associated Examining Board—Syllabus I)

6 (*a*) Write equations for the reactions occurring and name the products obtained at each electrode when an electric current is passed through an aqueous solution of copper(II) sulphate using **copper** electrodes. (4 lines, 4 marks)

(*b*) An electric current was passed through an aqueous solution of copper(II) sulphate using **carbon** electrodes. A sample of the colourless gas evolved at the positive electrode formed a white suspension when shaken with limewater. A second sample, after first shaking with aqueous sodium hydroxide, re-kindled a glowing splint. Explain these observations, giving relevant equations.

(*c*) An electric current of 0.40 A was passed through molten lead bromide for 40 minutes (using carbon electrodes). The apparatus was allowed to cool and the solid bead of lead metal formed at the cathode was removed and weighed. The mass of lead formed was 1.03 g. Use these results to show that the charge on the lead ion is 2 +. (Relative atomic mass of lead = 206, Faraday Constant F = 96 000 c/mol) (7 lines, 10 marks)

(Associated Examining Board—Syllabus I)

7 The equation for the combustion of ethanol is

$C_2H_5OH + 3O_2 \rightarrow 2CO_2 + 3H_2O; \Delta H = -1371$ kJ/mol

92 g of ethanol were completely burned in oxygen.
(Relative atomic masses C = 12, H = I, O = 16)

(*a*) How many moles of oxygen would be needed? (1 line, 2 marks)
(*b*) What volume of carbon dioxide would be formed at room temperature and pressure? (1 line, 2 marks)

(The volume of 1 mol of gas at room temperature is 24 dm^3)

(*c*) What quantity of energy would be released to the surroundings? (1 line, 2 marks)

(Associated Examining Board—Syllabus I)

8 (*a*) What do you understand by the term *oxidation*?
For each of the following reactions, state which substance is being oxidised and give a reason for your answer.

(*i*) $PbO + CO \rightarrow Pb + CO_2$
(*ii*) $2I^- + Br_2 \rightarrow I_2 + 2Br^-$
(*iii*) $2Fe^{3+} + H_2S \rightarrow 2Fe^{2+} + 2H^+ + S$
(*iv*) $Cu^{2+} + Zn \rightarrow Cu + Zn^{2+}$ (10 lines, 10 marks)

(*b*) When an electric current is passed between platinum electrodes immersed in aqueous copper(II) sulphate, what would you **see** happening at each electrode?

If the direction of current flow is now reversed and an identical quantity of electricity is passed through the solution, state what would be the final observation.

(Assume that the solution remains blue in colour throughout both experiments.)

Write ionic equations for any reactions occurring at the electrodes and indicate whether oxidation or reduction has taken place. (10 lines, 12 marks)

(*c*) When the same electric current was passed for a fixed time through aqueous copper(II) sulphate and aqueous mercury(I) nitrate, $HgNO_3$, in separate vessels equipped with suitable electrodes, it was found that 1.0 g of mercury was deposited. What mass, in g, of copper would also have been obtained? (6 lines, 4 marks)

(Cu = 64, Hg = 200)

(Associated Examining Board—Syllabus I)

9 (*a*) (*i*) Starting with aqueous sulphuric acid and aqueous sodium hydroxide, describe **in detail** how you would prepare pure, dry crystals of sodium sulphate. (Diagrams are **not** required.)

(*ii*) State **briefly** how you would modify your method in order to prepare a different sodium salt of sulphuric acid.

(*iii*) Write molecular equations for the reactions that occur in the two experiments. (12 lines, 17 marks)

(*b*) 0.61 g of metal hydroxide, $M(OH)_2$, exactly neutralises 20.0 cm^3 of a solution of hydrochloric acid containing 18.25 g of hydrogen chloride per dm^3. Write an equation for the reaction and calculate the relative molecular mass of the metal hydroxide; hence determine the relative atomic mass of M. (8 lines, 7 marks)

(H = 1, O = 16, Cl = 35.5)

(Associated Examining Board—Syllabus I)

Essay-type questions are becoming less popular on many examination papers and are usually restricted to 'O' level papers. Many examinations include essay questions that are structured to assist the candidates.

The following question is an example of a structured essay type of question. Before attempting the question, read each part of it through carefully. Read it through again and underline the important points. All too often a candidate does not answer the question being asked, but instead answers the question he *thinks* is being asked.

Divide up the time available so that each part of the question is attempted. Give more time to those parts of the question that carry most marks.

Before beginning an answer, make a rough plan in pencil of the main points to be included. When you have finished answering the question, cross out the plan so that the examiner does not try to mark it.

Your answer should be written in good English. The examiner is looking for specific points in your answer. There is no benefit, therefore, in filling out your answer to impress the examiner.

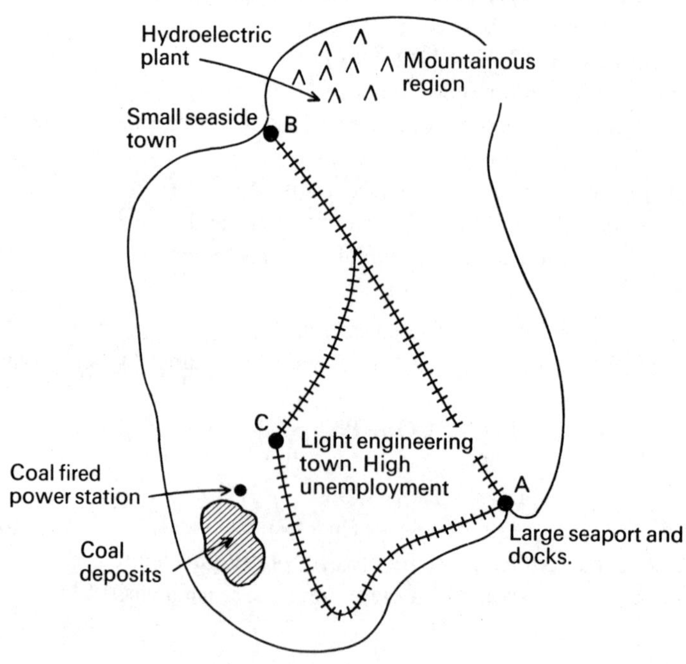

10 (*a*) Describe briefly how aluminium can be extracted from purified bauxite. Include in your answer a diagram of the plant and the equations. (10 lines, diagram, 8 marks)

(*b*) The map shows an imaginary island with no bauxite deposits. It is intended to set up a works to produce aluminium from imported bauxite to supply aluminium for this and neighbouring islands. Three sites labelled A, B and C have been discussed. List the advantages and disadvantages of each site. Which one of the three sites would you choose? (20 lines, 12 marks)

SCE questions (answers on page 190)

There is a new syllabus at 'O' grade from 1981. The following questions are reproduced, with permission from SEB, from the sample paper. In paper 1 of the Scottish 'O' grade examination, five marks out of the 50 (10%) are allocated to communication. This takes account of writing, spelling, grammar and punctuation.

Question with communication mark

11 (You are not expected to have encountered disodium dioxide before.)
Disodium dioxide is a white solid of formula Na_2O_2. When it is dropped into water, a gas is given off, the solid dissolves and an alkaline solution remains.

(*a*) Show, by means of a diagram, how you would try to collect a sample of the gas given off when disodium dioxide reacts with water. (diagram, 2 marks)

(*b*) Only TWO gases can possibly be given off in this reaction. Name them. (2 lines, 2 marks)

(*c*) How would you test your sample to decide which of these two gases you had collected? (2 lines, 1 mark)

(*d*) Suggest the name of the alkaline solution. (1 line, 1 mark)

(5 marks) (+ 1 mark for communication)

Questions without communication mark

12

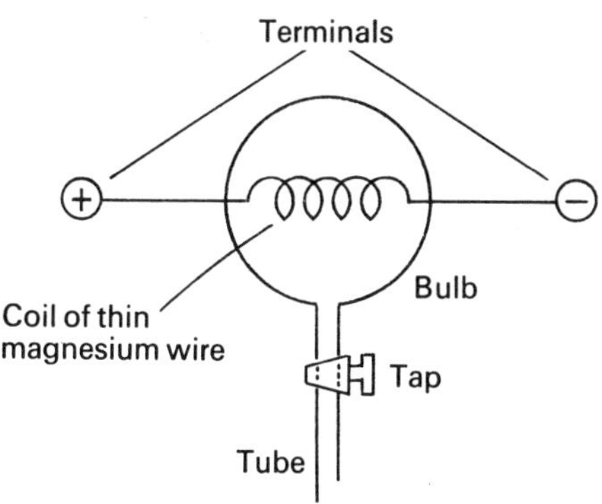

In the apparatus shown, the bulb is filled with oxygen, and the tap is closed. The magnesium is then heated by passing a current through it. There is a bright flash, and a white ash appears in the bulb.

(*a*) What is the white ash? (1 line, 1 mark)

(*b*) The tube is now dipped under water, and the tap is opened. What do you expect to happen, and why? (2 lines, 2 marks)

(*c*) If the original experiment is repeated using carbon dioxide instead of oxygen, the reaction is similar but the bulb becomes covered with black specks. Write a word equation for this reaction. (2 lines, 2 marks)

13 Cobalt chloride, $CoCl_2$, is a blue solid. Crystals of $CoCl_2.6H_2O$ are red. The solution is pink.

(*a*) What would be the most likely colour for $CoSO_4.6H_2O$? (1 line, 1 mark)

(*b*) Name another substance which changes colour when water of crystallisation is removed. (1 line, 1 mark)

(*c*) A colourless liquid was believed to be water. When a few drops were added to the blue solid $CoCl_2$ the colour changed to pink. Explain why this would not prove that the colourless liquid was water and describe how you would solve the problem. (2 lines, 3 marks)

14 The flow-diagram below shows part of a process used to make cement.

CALCIUM CARBONATE $\xrightarrow{\text{heat}}$ CALCIUM OXIDE $\xrightarrow{\text{x}}$ CALCIUM HYDROXIDE $\xrightarrow[\text{+ heat}]{\text{clay}}$ CEMENT

(*a*) What is the name of substance x? (1 line, 1 mark)

(*b*) Calcium, carbon, oxygen and hydrogen are present in compounds shown in the flow-diagram. Name one other element which is a major constituent of cement. (1 line, 1 mark)

(*c*) Name one other substance obtained in the first stage of the process but which is not shown on the flow-diagram. (1 line, 1 mark)

Answers to GCE and SCE questions

If you do not fully understand the answers given here, refer back to the units on which the questions are based.

1 (*a*) magnesium oxide or calcium oxide (Unit 5.6); (*b*) oxygen (Unit 32.9); (*c*) dinitrogen monoxide (nitrous oxide) not oxygen as oxygen is not a compound (Unit 29); (*d*) sodium hydrogencarbonate (Unit 19.9).

2 (*a*) A is copper(II) oxide.

$CuO + H_2SO_4 \rightarrow CuSO_4 + H_2O$ copper(II) sulphate—blue solution.

$CuSO_4 + 2NH_4OH \rightarrow Cu(OH)_2 + (NH_4)_2SO_4$ copper(II) hydroxide—pale blue precipitate (Unit 41)

Copper(II) hydroxide precipitate re-dissolves in excess ammonia solution (see unit 41.3)

(*b*) B is white phosphorus.

$P_4 + 5O_2 \rightarrow P_4O_{10}$ phosphorus(V) oxide

Phosphorus(V) oxide dissolves in water to form phosphoric acid.

$P_4O_{10} + 6H_2O \rightarrow 4H_3PO_4$

(Unit 31.3)

(*c*) C is calcium chloride. Brick red flame test confirms calcium (see unit 41.3). C is deliquescent. White precipitate of silver chloride confirms chloride.

$CaCl_2 + 2AgNO_3 \rightarrow 2AgCl + Ca(NO_3)_2$

or ionic equation $Cl^- + Ag^+ \rightarrow AgCl$

(*d*) D is sodium nitrate. It evolves oxygen on heating.

$2NaNO_3 \rightarrow 2NaNO_2 + O_2$ Sodium nitrite produced.

When warmed with concentrated sulphuric acid, nitric acid vapour produced.

$NaNO_3 + H_2SO_4 \rightarrow HNO_3 + NaHSO_4$

(Units 28.2 and 28.6)

3 (*a*) Na 2,8,1; F 2,7 (Unit 4.1).
(*b*) $Na \rightarrow Na^+ + e^-$ (Unit 5.1);

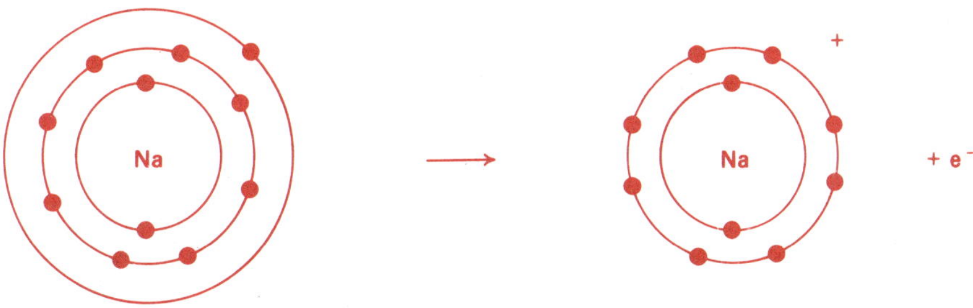

$F + e^- \rightarrow F^-$;

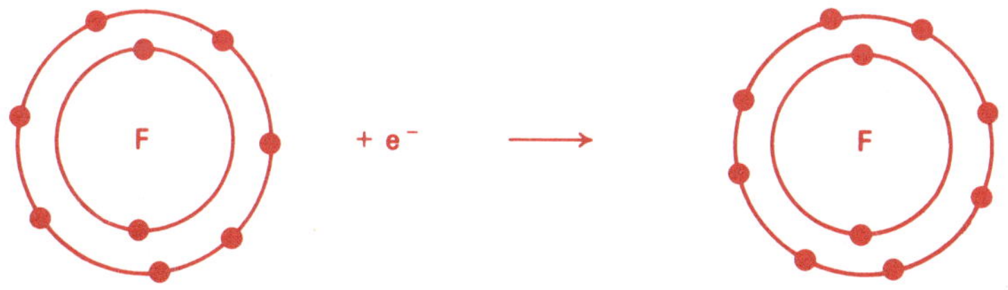

(*c*) High melting point, soluble in water (Unit 5.1).
(*d*) (*i*) Same electronic arrangements; (*ii*) different numbers of protons (Unit 5.1).
(*e*)

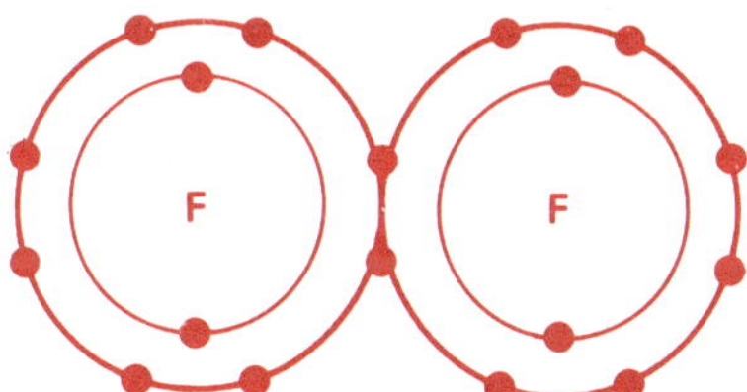

(Unit 5.1)

(*f*) (*i*) Isotopes–different numbers of neutrons;
(*ii*) Naturally-occurring neon consists of a mixture of isotopes, the main constituent being neon-20. (Unit 4.1)

4 (*a*) Lead(II) sulphate is an insoluble salt and has to be prepared by precipitation (see unit 19.13). The first step is to convert lead(II) oxide to lead(II) nitrate (a soluble salt).

$$PbO + 2HNO_3 \rightarrow Pb(NO_3)_2 + H_2O$$

Lead(II) oxide is warmed with dilute nitric acid. Excess lead(II) oxide is filtered off. Dilute sulphuric acid is added to lead(II) nitrate solution and the mixture is stirred.

$$Pb(NO_3)_2 + H_2SO_4 \rightarrow PbSO_4 + 2HNO_3$$

Lead sulphate precipitate is removed by filtration, washed with water and dried.

(*b*) Copper carbonate heated. $CuCO_3 \rightarrow CuO + CO_2$
Copper(II) oxide reduced with hydrogen (see Unit 10.5). $CuO + H_2 \rightarrow Cu + H_2O$

(*c*) Calcium carbonate heated strongly. $CaCO_3 \rightarrow CaO + CO_2$
Carbon dioxide is passed over heated carbon. $CO_2 + C \rightarrow 2CO$ (Unit 34.5)

5 (*a*) (*i*) A 32 g; B 48 g.
(*ii*) These masses combining with a fixed mass of molybdenum are in a simple ratio, *i.e.* 2:3. This verifies the Law of Multiple Proportions.
(*iii*) Oxide A: 1.92 g of Mo combines with 0.64 g of oxygen
192 g of Mo combines with 64 g of oxygen
$\frac{192}{96}$ moles of Mo combines with $\frac{64}{16}$ moles of oxygen.
Simplest formula MO_2
Oxide B: simplest formula M_2O_3.

(Unit 13.3)

(*b*) (*i*) (see unit 24.8); (*ii*) See question 4(c).

6 (*a*) Cathode $Cu^{2+} + 2e \rightarrow Cu$ copper
Anode $Cu \rightarrow Cu^{2+} + 2e$ copper(II) ions formed (Unit 14.10)

(*b*) Electrolysis of copper(II) sulphate solution using carbon electrodes produces oxygen at the anode.

$$4OH^- \rightarrow 2H_2O + O_2$$

The carbon electrode is not completely inert and the gas produced is carbon dioxide which turns limewater milky.

$$C + O_2 \rightarrow CO_2$$
$$Ca(OH)_2 + CO_2 \rightarrow CaCO_3 + H_2O$$

(Unit 33.3)

(*c*) A current of 0.40 A was passed for 40 minutes.

Number of coulombs passed $= 0.40 \times 40 \times 60$
$= 960$ coulombs

Number of Faradays passed $= \frac{960}{96000}$
$= 0.01$ Faradays

Mass of lead produced by 0.01 Faradays $= 1.03$ g
Mass of lead produced by 1 Faraday $= 1.03 \times 100$
$= 103$ g

Relative atomic mass of lead $= 206$

Since 1 Faraday produces 0.5 moles of lead atoms and therefore
2 Faradays produce 1 mole of lead atoms,
the charge on the lead ion is $2+$.

(Units 17.6 and 17.7)

7 (*a*) 92 g of ethanol $= \frac{92}{2 \times 12 + 6 \times 1 + 1 \times 16} = \frac{92}{46} = 2$ moles

Number of moles of oxygen molecules required from the equation to completely burn 2 moles of ethanol is 6.

(*b*) Number of moles of carbon dioxide produced from the equation $= 4$.
Since the volume of 1 mole of any gas at room temperature and pressure is 24 dm^3, volume of carbon dioxide $= 4 \times 24\ dm^3 = 96\ dm^3$

(*c*) Quantity of energy released $= 2 \times 1371$kJ
$= 2742$kJ (Units 12.10 and 37)

8 (*a*) Oxidation is a reaction where oxygen is added, hydrogen or electrons are lost.
(*i*) carbon monoxide (CO) is oxidised to carbon dioxide because it gains oxygen.
(*ii*) iodide ions (I^-) are oxidised to iodine molecules (I_2) because electrons are lost $2I^- \rightarrow I_2 + 2e$.
(*iii*) hydrogen sulphide (H_2S) is oxidised to sulphur (S) because hydrogen is lost.
(*iv*) zinc atoms (Zn) are oxidised to zinc ions (Zn^{2+}) because electrons are lost $Zn \rightarrow Zn^{2+} + 2e$.

(Unit 16.1)

(*b*) It is important to explain what is *seen* at each electrode. The electrodes are inert like carbon and are therefore not involved.
Positive electrode (Anode)
Bubbles of colourless oxygen gas are evolved from the electrode surface.

$$4OH^- \rightarrow 2H_2O + O_2 + 4e$$

OH^- ions lose electrons and are therefore oxidised.
Negative electrode (Cathode)
Brown copper is deposited on the platinum electrode and a *few* bubbles of colourless hydrogen gas may be produced. The colour of the solution around the electrode may fade.

$$Cu^{2+} + 2e \rightarrow Cu$$

Cu^{2+} ions gain electrons and are therefore reduced. (Unit 17.3)

N.B. The process at the anode is always oxidation and the process at the cathode is always reduction.
If the current is reversed, the copper deposited on the electrode goes into solution and no gaseous product is formed.

$$Cu \rightarrow Cu^{2+} + 2e \quad \text{oxidation}$$

(*c*) Mercury(I) nitrate contains Hg^{+} ions and copper(II) sulphate contains Cu^{2+} ions. The same quantity of electricity will deposit twice the number of moles of mercury atoms.

Number of moles of mercury atoms deposited $= \frac{1}{200}$ moles

$\therefore$ Number of moles of copper atoms deposited $= \frac{1}{400}$ moles

Mass of copper deposited $= \frac{1}{400} \times 64$ g

$= 0.16$ g (Units 12.5, 12.6, 17.6 and 17.7)

9 (*a*) (*i*) Pipette 25 cm^3 (or other suitable volume) of dilute sodium hydroxide solution into a conical flask. Add a few drops of a suitable indicator (e.g. screened methyl orange). Add the dilute sulphuric acid in small quantities from a burette, shaking after each addition until the solution goes colourless. Note the volume of dilute sulphuric acid added.
Repeat the experiment with a fresh sample of sodium hydroxide solution. Do not add the indicator but add the same volume of dilute sulphuric acid as before.
Evaporate the solution in an evaporating basin until a small volume of solution remains. The solution is then left to cool and crystallise. Finally the crystals of sodium sulphate can be dried between filter papers.

(*ii*) Repeat the experiment as in (*i*) but use 12.5 cm^3 of sodium hydroxide solution and the same volume of dilute sulphuric acid (see (*iii*)).

(*iii*) $2NaOH(aq)^{+} + H_2SO_4(aq) \rightarrow Na_2SO_4(aq) + 2H_2O(l)$
sodium hydroxide + sulphuric acid → sodium sulphate + water
$NaOH(aq) + H_2SO_4(aq) \rightarrow NaHSO_4(aq) + H_2O(l)$
sodium hydroxide + sulphuric acid → sodium hydrogensulphate + water
ACID SALT

(Units 19.2 and 40)

(*b*) $M(OH)_2 + 2HCl \rightarrow MCl_2 + 2H_2O$
20.0 cm^3 of dilute hydrochloric acid neutralises 0.61 g $M(OH)_2$
$\therefore$ 2000 cm^3 of dilute hydrochloric acid neutralises 61 g $M(OH)_2$
2000 cm^3 of dilute hydrochloric acid (i.e. 2 dm^3) contains 36.5 g of HCl i.e. 1 mole of hydrochloric acid
Relative molecular mass of the metal hydroxide = 122 g (from the equation, it is the mass which will react with 2 moles of hydrochloric acid).
Relative atomic mass of M = Relative molecular mass of $M(OH)_2$
− (2 × relative atomic mass of O)
− (2 × relative atomic mass of H)
= 122 − 32 − 2 g
= 88 g

(Unit 40)

10 (*a*) See unit 15.6.

(*b*) *The following answer is the skeleton on which to hang your essay. Make sure that you write in complete sentences and that you express your arguments as logically as possible.*

Site A. Good port facilities for import of bauxite and export of finished aluminium. Electricity could be transported from hydroelectric plant or coal-fired power station.

Site B. Bauxite and finished aluminium would have to be transported by rail. Electricity from hydroelectric plant readily available. Possible difficulties with availability of labour and environmental complaints.

Site C. Bauxite and finished aluminium transported by rail. Possible to use some of the aluminium in local industry. Electricity available from coal-fired power station and labour readily available.

On balance, site A is probably best. It is easier to transport electricity than bauxite or aluminium. If there were problems with labour or exact siting, site C would be possible.

(Unit 38)

11 (*a*) Fig. 26.2 shows an experiment with suitable apparatus.
(*b*) Oxygen and hydrogen (Unit 8).
(*c*) Test the gas with a burning splint. If the gas burns with a squeaky pop the gas is hydrogen. With oxygen the splint will burn brighter. If a glowing splint is put into oxygen the splint relights (Unit 8).
(*d*) Sodium hydroxide.

12 (*a*) Magnesium oxide.
(*b*) Water would enter the bulb and completely fill it. The oxygen inside the bulb has been used up.

$$2Mg + O_2 \rightarrow 2MgO$$

(*c*) carbon dioxide + magnesium   magnesium oxide + carbon

(Units 8 and 9.2)

13 (*a*) Pink.
(*b*) Hydrated copper(II) sulphate $CuSO_4.5H_2O$.
(*c*) Cobalt chloride turning pink shows that water is present but any aqueous solution would do the same. Pure water (at atmospheric pressure) boils at 100°C (Unit 11.6 and 11.8).

14 (*a*) Water.
(*b*) Silicon.
(*c*) Carbon dioxide.

(Unit 14.4)

CSE questions (answers on page 195)

The first two questions are factual and the answers should consist of a single word or phrase.

1 Complete the following table:

Element	*Symbol*	*Liquid, solid or gas*
Lithium	Li	Solid
	I	
	Pb	Solid
Potassium		

(East Anglian Examinations Board—South)

2 Complete the following:
When iron rusts it combines with.........and.........from the air and forms.........

When scrap iron is thrown into streams and canals, pollution is caused by the metal removing the gas.........that is dissolved in the water.

(West Midlands Examinations Board—Syllabuses A and B)

Questions 3–10 are structured questions. A certain amount of information is given at the beginning of each question followed by a series of short questions which are related to the information given. First read thoroughly and understand the information given. Then look at the amount of space provided (if the question is to be fitted into spaces on the question paper) or at the maximum marks to be awarded (often shown in brackets after the question). These are clues to how much information is required.

3 A student set up the apparatus shown below in order to investigate the products formed when a fraction of crude oil burns.

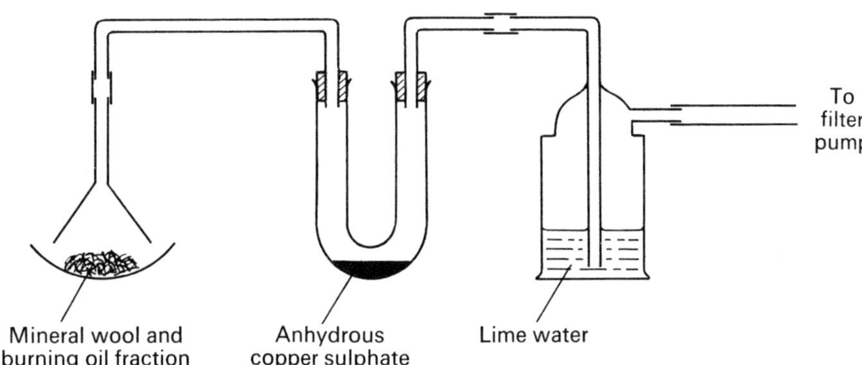

(*a*) What is the purpose of the filter pump in this experiment? (1 line, 1 mark)
(*b*) What change would occur in the limewater after a short time? (1 line, 1 mark)
Name the substance which causes this change. (1 line, 1 mark)
(*c*) What change would occur in the anhydrous copper sulphate after a short time? (2 lines, 2 marks)
Name the substance which causes this change. (1 line, 1 mark)
(*d*) What colour change is seen around the inside of the filter funnel? (1 line, 1 mark)
(*e*) Using the results of the experiment, name two elements present in the oil. (1 line, 2 marks)
(*f*) Another student set up a similar apparatus with the 'U' tube and the wash bottle interchanged. Why did his teacher point out that this was unsuitable for this experiment? (3 lines, 4 marks)
(*g*) What difference would it make if the experiment was left running for the same length of time without the burning oil? (2 lines, 2 marks)
What does this second experiment prove?
(*h*) Assuming that some of the oil was pentane (C_5H_{12}) complete and balance the equation for its burning assuming complete combustion.

$$C_5H_{12} + O_2 \rightarrow$$ (1 line, 2 marks)

(*i*) Name another substance which would have given the same results in the experiment as the oil fraction. (1 line, 1 mark)

(Associated Lancashire Schools Examining Board)

4 Three gaseous oxides of nitrogen have the formulae

NO	N_2O	NO_2
(1)	(2)	(3)

(*a*) How many oxygen atoms are there in one molecule of N_2O? (1 line, 1 mark)
(*b*) Which nitrogen oxide contains the highest proportion of nitrogen? (1 line, 1 mark)
(*c*) When oxide (1) is passed over heated copper turnings, copper(II) oxide and nitrogen are formed. Write the chemical equation for this reaction. (1 line, 2 marks)
(*d*) Oxide (1) combines with oxygen according to the following equation:

$$2NO + O_2 = 2NO_2$$

(*i*) How many cubic centimetres of oxygen would combine with 10 cm^3 of oxide (1) and how many cubic centimetres of oxide (3) would be formed?

.... cm^3 of oxygen

.... cm^3 of oxide (3) (2 lines, 3 marks)

(*ii*) How many grammes of oxide (3) could be made from 3 g of oxide (1)? (3 lines, 2 marks)

(N = 14, O = 16)

(East Anglian Examinations Board—South)

5 (*a*) Write down the formula of sulphuric acid. (1 line, 1 mark)

(*b*) (*i*) Describe a test to show that a given liquid is an acid. (4 lines, 1 mark)

(*ii*) Describe a test to show a given acid is dilute sulphuric acid. (2 lines, 2 marks)

(*c*) Using dilute sulphuric acid, describe chemical tests you might use to distinguish between the substances in each of the following pairs. Describe also how you would identify any gases given off.

(*i*) Sodium sulphite and sodium hydrogencarbonate (sodium bicarbonate) (5 lines, 4 marks)

(*ii*) Copper(II) oxide and powdered carbon (4 lines, 4 marks)

(Yorkshire Regional Examinations Board)

6 An experiment was carried out to investigate the reaction between copper carbonate and hydrochloric acid. 0.1 g of copper carbonate was weighed out on to a piece of paper and, as it was dropped into 50 cm^3 of 0.1M hydrochloric acid, a stopclock was started. The time was measured for the copper carbonate to dissolve completely and the effervescence to cease.

This was then repeated, but with different molarities of hydrochloric acid, as shown in the table below.

Volume of acid (cm^3)	*Molarity of acid*	*Weight of copper carbonate*	*Time to dissolve (seconds)*
50	0.1 M	0.1 g	75
50	0.2 M	0.1 g	70
50	0.5 M	0.1 g	50
50	1.0 M	0.1 g	30
50	2.0 M	0.1 g	20

(*a*) Which was the least concentrated acid? (1 line, 1 mark)

(*b*) What effect did the concentration of the acid have on the reaction? (2 lines, 1 mark)

(*c*) Name the gas which was evolved when the copper carbonate reacted with dilute hydrochloric acid. (1 line, 1 mark)

(*d*) Suggest a way by which the time for the complete reaction could be shortened without changing the concentration of the acid. (2 lines, 1 mark)

(*e*) If you started this experiment with a bottle of the 2.0M hydrochloric acid, how would you prepare 100 cm^3 of 0.1M hydrochloric acid from it? (2 lines, 1 mark)

(East Midland Regional Examinations Board—Syllabus I)

7 Magnesium nitride is a solid compound formed when magnesium is strongly heated in dry nitrogen gas until a reaction takes place. In an experiment it was found that 0.72 g of magnesium produced 1.00 g of magnesium nitride.

(*a*) Magnesium nitride is also formed when magnesium burns in dry air. Name the other compound formed when magnesium burns in dry air. (1 line, 1 mark)

(*b*) How many moles of magnesium atoms are present in 0.72 g of magnesium ribbon? (Mg = 24) (1 line, 1 mark)

(*c*) Calculate the mass of nitrogen which combined with 0.72 g of magnesium in this experiment. (1 line, 1 mark)

(*d*) How many moles of nitrogen atoms combined with 0.72 g of magnesium? (N = 14)
(1 line, 1 mark)

(*e*) How many moles of magnesium atoms would combine with two moles of nitrogen atoms?
(1 line, 1 mark)

(*f*) What is the simplest formula for magnesium nitride? (1 line, 1 mark)

When magnesium nitride is added to hot water a colourless gas X is produced. 40 cm^3 of X, which is a compound, was passed over heated iron wool to decompose it completely into two elements. The volume of the mixture of elements was 80 cm^3.

After passing the mixture over heated copper(II) oxide, the final volume was 20 cm^3. This gas was nitrogen (N_2).

(All volumes were measured at room temperature and pressure.)

(*g*) When X was decomposed, the other element produced was hydrogen (H_2).

(*i*) What volume of hydrogen was present in the 80 cm^3 sample of the mixture of element?
(1 line, 1 mark)

(*ii*) Explain, in words or with an equation, the reaction that takes place when the mixture is passed over heated copper(II) oxide. (2 lines, 2 marks)

(*h*) What is the formula of the gas X? (1 line, 1 mark)

(Southern Regional Examinations Board R(B))

8 The letters A, B, C, D, E, F and G represent seven elements. (These letters are not the actual symbols for these elements.) You are not expected to try to identify any of these elements.

	Atomic number	Melting point °C	Boiling point °C	Element heated in air	pH of oxide residue	Reaction of element with water
A	82	327	1744	Slow reaction	7	None
B	19	64	760	Burns on heating	12	Violent reaction
C	12	650	1110	Burns on heating	10	Reacts very slowly with cold water but rapidly with steam
D	3	180	1330	Burns on heating	11	Reacts steadily with cold water
E	80	−39	357	Slow reaction over several days	7	None
F	11	98	890	Burns on heating	11	Fast reaction
G	16	119	445	Burns on heating	5	None

Use the information above to answer the following questions. In each question give the letter or letters required and a brief reason for your answer.

(*a*) Which element is liquid at room temperature (20°C)? (1 line, 1 mark)

(*b*) Which element is most reactive? (1 line, 1 mark)

(*c*) Give one element that forms a neutral oxide. (1 line, 1 mark)

(*d*) Which two elements are adjacent (next to each other) in a period of the Periodic Table?
(1 line, 1 mark)

(*e*) Which three elements are in the same group of the Periodic Table? (1 line, 1 mark)

(*f*) Which element is a non metal? (1 line, 1 mark)

(*g*) What is the structure of element G at room temperature? (1 line, 1 mark)

If samples of A are put into solutions of the nitrates of C and E, a reaction takes place with the nitrate of E but not with the nitrate of C.

(*h*) Using all the information in the question, arrange the four elements A, C, D and E in order of reactivity, with the most reactive first. (3 lines, 2 marks)

(Southern Regional Examinations Board R (B))

9 The following table shows part of the Periodic Table with only a few symbols included.

GROUP / PERIOD	I	II											III	IV	V	VI	VII	O
1																		He
2																	F	
3	Na												Al	Si		S		
4		Ca				Cr											Br	
5																		
6																		

(*a*) Using only the elements in the table write down the symbol for each of the following:

(*i*) A metal stored in oil. (1 line, 1 mark)

(*ii*) A noble or inert gas. (1 line, 1 mark)

(*iii*) A transition metal. (1 line, 1 mark)

(*iv*) A halogen which is a gas at room temperature and pressure. (1 line, 1 mark)

(*v*) The element with the largest number of protons in the nucleus of each atom. (1 line, 1 mark)

(*vi*) The element which would contain most atoms in a 100 g sample. (1 line, 1 mark)

(*vii*) An element extracted from sand when a mixture of dry sand and magnesium powder is heated. (1 line, 1 mark)

(*b*) When Mendeleef devised the Periodic Table many of the elements that are now known had not been discovered. The following account refers to such an element. It is represented by the symbol X but this is not the usual symbol for the element.

X is an element with a melting point of 30°C and a boiling point of 2440°C. It conducts electricity at room temperature.

It burns in oxygen to form an oxide which has a pH of 7. The oxide is a colourless solid with a formula X_2O_3.

X forms a compound with fluorine which is a high melting point solid that conducts electricity when molten. The similar compound with bromine has a formula XBr_3 but is a low melting point solid.

The approximate relative atomic mass of X is 70.

(*i*) Is X a solid, liquid or gas at room temperature (20°C)? (1 line, 1 mark)

(*ii*) Give a reason why X is regarded as a metal. (2 lines, 1 mark)

(*iii*) Predict the formula for the fluoride of X. (1 line, 1 mark)

(*iv*) What are the products of the electrolysis of the molten fluoride of X? (2 lines, 2 marks)

(*v*) Give the symbol for the element in the above table which most closely resembles X. (1 line, 1 mark)

(*vi*) In which group and period of the Periodic Table will X be placed? (1 line, 1 mark)

(East Anglian Examinations Board)

10 (*a*) Some powdered starch was mixed thoroughly with dry copper(II) oxide and the mixture was heated in a dry test tube. Droplets of a colourless liquid collected on the cooler part of the test tube and carbon dioxide was produced.

The residue in the test tube at the end of the experiment includes a brown solid.

(*i*) The colourless liquid produced is water. Describe a test that could be used to show this. (1 line, 1 mark)

(*ii*) Identify the brown solid present in the residue. (1 line, 1 mark)

(*iii*) What is the reason for adding copper(II) oxide to the starch before heating? (2 lines, 1 mark)

(*iv*) What can be concluded about starch as a result of this experiment? (1 line, 1 mark) (*b*) An aqueous solution of starch was divided into four portions. These were used for four separate experiments.

(*i*) A few drops of iodine solution were added. A dark blue-black colouration was formed.

(*ii*) Starch solution and dilute hydrochloric acid were mixed and boiled. A pale yellow solution was produced when iodine solution was added.

(*iii*) Saliva was added to starch solution. A pale yellow solution was again produced when iodine solution was added.

(*iv*) Yeast was added to starch solution and the mixture kept in a warm place. The solution started to froth and a colourless gas was produced. Ethanol (boiling point 78°C) was present in the solution at the end of the experiment.

1. What is meant by the term 'aqueous solution'? (1 line, 1 mark)
2. What can be concluded from experiments (*ii*) and (*iii*)?
3. At the end of experiment (*ii*) a reducing sugar is present in the solution.
   a Which reagent could be used to test for a reducing sugar? (1 line, 1 mark)
   b What colour change would be observed when this reagent is heated with the solution from experiment (*ii*)? (1 line, 1 mark)
4. What name is given to the process occurring in experiment (*iv*)? (1 line, 1 mark)
5. Name the colourless gas evolved during experiment (*iv*). (1 line, 1 mark)
6. What is present in yeast which produces the reaction in experiment (*iv*)? (1 line, 1 mark)
7. Draw a labelled diagram of apparatus that could be used in order to obtain a fairly pure sample of ethanol from the solution remaining from experiment (*iv*). Give the name of this process. (diagram, 3 marks)

(East Anglian Examinations Board)

Answers to CSE questions

If you do not fully understand the answers given here, refer back to the units on which the questions are based.

1 Iodine, solid; lead; K, solid. (Unit 3)

2 Oxygen (or water vapour), water vapour (or oxygen), hydrated iron oxide or iron oxide, oxygen. (Unit 8.6)

3 (*a*) To draw air and the products of combustion through the apparatus.
(*b*) Limewater would turn milky—carbon dioxide.
(*c*) White anhydrous copper sulphate turns blue—water.
(*d*) Turns black.
(*e*) Carbon, hydrogen (oxygen could also be present).
(*f*) Water present in limewater would turn anhydrous copper sulphate blue.
(*g*) In the same time, limewater would not turn milky. There is much less carbon dioxide in air and so carbon dioxide must be produced from oil burning.
(*h*) $C_5H_{12} + 8O_2 \rightarrow 5CO_2 + 6H_2O$
(*i*) Range of answers including candle wax, methane, ethanol *etc.*
(Units 11.8, 34.3 and 33.15)

4 (*a*) 1; (*b*) N_2O; (*c*) $2NO + 2Cu \rightarrow 2CuO + N_2$; (*d*) (*i*) 5 cm^3; 10 cm^3; (*ii*) 4.6 g. (Unit 12)

5 (*a*) H_2SO_4; (*b*) (*i*) (see Unit 19.2); (*ii*) dilute hydrochloric acid and barium chloride solution (see unit 41.2).
(*c*) (*i*) Sodium sulphite produces sulphur dioxide on warming. This turns filter paper soaked in potassium dichromate solution from orange to green. Sodium hydrogencarbonate evolves carbon dioxide without heating. No smell.
(*ii*) Copper(II) oxide reacts on warming to produce blue copper(II) sulphate solution. Carbon does not react with dilute sulphuric acid.

$$CuO + H_2SO_4 \rightarrow CuSO_4 + H_2O$$ (Unit 41)

6 (*a*) 0.1 M; (*b*) Increasing concentration speeds up reaction (decreases time); (*c*) CO_2; (*d*) Heating the acid; (*e*) Measuring out exactly 5 cm^3 of 2 M hydrochloric acid and making up to 100 cm^3. (Unit 20)

7 (*a*) Magnesium oxide
(*b*) 0.03 moles, *i.e.* 0.72 ÷ 24
(*c*) 1.00 g − 0.72 g = 0.28 g
(*d*) 0.02 moles, *i.e.* 0.28 ÷ 14
(*e*) 3
(*f*) Mg_3N_2
(*g*) (*i*) 60 cm^3 (*ii*) copper(II) oxide + hydrogen → copper + water
$$CuO + H_2 \rightarrow Cu + H_2O$$
(*h*) NH_3

8 (*a*) E. Room temperature between melting and boiling points.
(*b*) B. Violent reaction with water or most alkaline oxide.
(*c*) A or E. pH of oxide is 7.
(*d*) F and C. Only one different in atomic number.
(*e*) D, F and B are members of the alkali metal family. Relationship is between atomic numbers, boiling and melting points.
(*f*) G. Forms an acidic oxide.
(*g*) Molecular structure—low melting point.
(*h*) D,C,A,E. (Units 15.10 and 18)

9 (*a*) (*i*) Na; (*ii*) He; (*iii*) Cr; (*iv*) F; (*v*) Br; (*vi*) He; (*vii*) Si.
(*b*) (*i*) Solid.
(*ii*) One of the following—oxide neutral, forms positive ions, conducts electricity when solid.
(*iii*) XF_3.
(*iv*) Positive electrode fluorine. Negative electrode X.
(*v*) Al.
(*vi*) Group III Period 4. (Unit 18)

10 (*a*) (*i*) To show that the liquid is water it should be heated and a boiling point of 100°C obtained. Anhydrous copper(II) sulphate turning from white to blue or cobalt(II) chloride paper turning from blue to pink are tests which show that a liquid contains water.
(*ii*) Copper.
(*iii*) Supplying oxygen for combustion.
(*iv*) Starch contains carbon and hydrogen.
(*b*) 1. See glossary p 197.
2. Starch has reacted or hydrolysis has taken place.
3. a Benedict's solution or Fehling's solution.
b Blue or green changes to yellow/orange/red.
4. Fermentation.
5. Carbon dioxide.
6. Enzymes.
7. Refer to Fig. 2.4.
Fractional distillation.

(Units 33.26—33.28)

Glossary

The words in this glossary are words commonly used in Chemistry. Teachers and examiners may assume you understand the meanings of these words. If you find a word you do not understand in this book or in an examination paper look it up in this glossary. If the word does not appear in the glossary it will be explained within the book and you should refer to the index. For example the word 'catalyst' appears a number of times in this book. It does not appear in the glossary because it is explained on page 78.

Absolute temperature. There is a minimum temperature below which it will never be possible to cool anything. This is called the **absolute zero** and is −273°C. This is the starting point for the Kelvin, or absolute temperature, scale: *e.g.* 0°C is the same as 273 K which is called the absolute temperature.

Amalgams. Many metals form (with mercury) alloys called amalgams. The mixture used to fill teeth is an amalgam.

Anhydride. An anhydride (sometimes called an acid anhydride) is an oxide of a non-metal which dissolves in water to form an acid.

Aqueous. An aqueous solution is a solution in which the solvent is water.

Condensing. When vapour is cooled it turns to a liquid and gives out heat. This change is called condensing.

Corrosion. The effect of air and water vapour on metals is called corrosion. Rusting of iron is the most common example of corrosion.

Crystal. A crystal is a regularly shaped solid. Within the solid the constituent particles are regularly arranged.

Decomposition. When a substance decomposes it splits up into new substances. If the splitting up requires heat it may be called **thermal decomposition**. If the substance reforms the original substance on cooling, *e.g.* ammonium chloride, it is called **dissociation**.

Dehydration. Dehydration is the removal of water. A **dehydrating agent**, such as concentrated sulphuric acid, is good at removing water.

Detergent. Any cleaning agent is called a detergent. The main types of detergent are soaps and soapless detergents.

Diatomic. An element whose molecules are composed of two atoms is said to be diatomic. The common gases oxygen, hydrogen, nitrogen and chlorine are all diatomic and are written as O_2, H_2, N_2 and Cl_2.

Discharge. An ion may be converted to a neutral atom at an electrode during electrolysis by loss or gain of electrons. This is called discharging of ions.

Ductile. A metal is said to be ductile as it can be drawn into thin wires.

Effervescence. If a gas is produced during a chemical reaction bubbles of gas can be seen to escape from the solution. This 'fizzing' is called effervescence and this word is frequently confused with **efflorescence**.

Evaporating. Evaporation is the change of a liquid into a vapour or gas. With water this takes place slowly at room temperature and more quickly at higher temperatures. Evaporation which takes place at a particular temperature, called the boiling point, is called **boiling**.

Excess. In a chemical reaction there is a connection between the quantities of substances reacting. In practice one of the reactants is present in larger quantities than is required for the reaction and is said to be in excess.

Freezing. This is the changing of a liquid into a solid. It is the opposite process to **melting**.

Hydrated. Hydrated solids, such as copper sulphate crystals ($CuSO_4.5H_2O$), contain a fixed amount of water within the crystal. This water is called **water of crystallisation** and can be lost on heating. The resulting solid is said to be **anhydrous**.

Hydrocarbons. Hydrocarbons are compounds of the elements carbon and hydrogen only.

Hydrolysis. Hydrolysis is the splitting up of a substance with water. Hydrolysis reactions are often slow with water but faster with dilute acids or alkalis.

In situ. It is sometimes necessary to prepare a chemical, as it is required, by mixing other chemicals. Ammonium nitrate, which is inclined to be explosive because of impurities, can be prepared by mixing ammonium chloride and sodium nitrate. The ammonium chloride is said to be prepared *in situ*.

Ion. When an atom or group of atoms gains or loses one or more electrons an ion is produced. It will be negatively or positively charged. This process is called **ionisation**.

Malleable. A metal can be beaten into thin sheets and is said to be malleable.

Miscible. Two liquids which mix completely are said to be miscible. Liquids which do not mix but form separate layers are said to be **immiscible**.

Platinised. A platinised asbestos catalyst is a piece of substitute asbestos covered with the thinnest coating of platinum. This will provide the largest possible surface area of platinum from a given mass of platinum.

Polar solvent. The molecules of some solvents, such as water, contain slight positive and negative charges. A stream of a polar solvent is deflected by a charged rod. A polar solvent dissolves substances containing ionic bonds. Solvents without these charges are called **non-polar solvents**.

Precipitate. If a solid which is insoluble in water is produced during a chemical reaction in solution, the particles of solid formed are called a precipitate and the process is called **precipitation.**

Proportional. If a car is driving along at a constant speed the amount of petrol used increases regularly as the distance increases. The amount of petrol used is said to be proportional to the distance travelled.

Qualitative. A qualitative study is one which depends upon changes in appearance only. A **quantitative** study requires a study of quantities, *e.g.* mass, volume, etc.

Reactants. The substances which are used at the start of a reaction are called reactants.

Receiver. The vessel in which the distillation is collected during distillation is called the receiver.

Refractory. A refractory is a high melting point ionic oxide used for lining furnaces.

Saturated. There are two uses of this word in Chemistry. A **saturated solution** is a solution in which the maximum amount of solute is dissolved at a particular temperature. A carbon compound which contains only single bonds between atoms is said to be saturated. If double and/or triple bonds are present it is said to be **unsaturated**.

Soluble. A substance which dissolves in a **solvent** is said to be soluble, and a substance which does not dissolve to be **insoluble**. Substances which are only slightly soluble are called **sparingly soluble**.

Solute. The substance which is dissolved in a **solvent** is called the solute. The resulting mixture is called a **solution**.

Vapour. A vapour is a gas which will condense to a liquid on cooling to room temperature.

Viscous. A viscous liquid is thick and 'treacle-like'. It is difficult to pour.

Volatile. A liquid, which is volatile such as petrol, boils at a low temperature and easily vaporises. An **involatile** liquid does not evaporate quickly.

Index